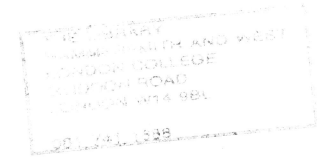

Nutritional Needs of Athletes

Nutritional Needs of Athletes

FRED BROUNS

Nutrition Research Center, University of Limburg, Maastricht,
The Netherlands

JOHN WILEY & SONS

Chichester · New York · Brisbane · Toronto · Singapore

Other Wiley Editorial Offices

John Wiley & Sons, Inc., 605 Third Avenue,
New York, NY 10158-0012, USA

Jacaranda Wiley Ltd, G.P.O. Box 859, Brisbane,
Queensland 4001, Australia

John Wiley & Sons (Canada) Ltd, 22 Worcester Road,
Rexdale, Ontario M9W 1L1, Canada

John Wiley & Sons (SEA) Pte Ltd, 37 Jalan Pemimpin #05-04,
Block B, Union Industrial Building, Singapore 2057

Library of Congress Cataloging-in-Publication Data

Brouns, F. (Fred).
    Nutritional needs of athletes / Fred Brouns.
        p.   cm.
    Includes bibliographical references and index.
    ISBN 0-471-94079-8 : £19.95
    1. Athletes—Nutrition.   I. Title.
    TX361.A8B76   1993
    613.2'024796—dc20
                                                    93-11818
                                                       CIP

British Library Cataloguing in Publication Data

A catalogue record for this book is available
from the British Library

ISBN 0 471 94079 8

Typeset in 11/13pt Palatino by
Mathematical Composition Setters Ltd, Salisbury, Wiltshire
Printed and bound in Great Britain by
Biddles Ltd, Guildford and King's Lynn

31515

Contents

Preface

The present manuscript is meant to give a scientific overview of aspects related to nutrition and physical activity, especially of people involved in intensive sports performance.

The manuscript is based on a large number of recent scientific reviews and publications, which have appeared in peer reviewed scientific journals. This means that these publications have generally survived the criticisms of the reviewers and that the interpretations are in line with existing scientific consensus.

To achieve a maximal degree of scientific consensus, the draft manuscript has been sent to a number of experts in the field of exercise science and nutrition. Selection of these experts was based on their actual research activities and their internationally known expertise in different fields of sport nutrition.

Their review and criticisms have resulted in a further improvement of the final manuscript, which is presented here.

Fred Brouns

Acknowledgements

The helpful contribution of the following experts who critically reviewed and discussed the draft text, to realize a status of scientific consensus in the final manuscript, is greatly acknowledged:

- Professor Dr Melvin Williams, Human Performance Laboratory, Old Dominion University, Norfolk, USA. Expert in sport nutrition science, Editor in chief of the *International Journal of Sports Nutrition*. A special word of thanks goes to Professor Williams for his inputs to the chapter on ergogenic acids.
- Professor Dr Ir Wim H. M. Saris, Nutrition Research Center, Department of Human Biology, University of Limburg, Maastricht, The Netherlands. Expert in human nutrition/ sport nutrition.
- Professor Dr Abel Mariné-Font, Universitat de Barcelona, Dep. de Ciències Fisiològiques Humanes i de la Nutrició, Unitat de Nutrició i Bromatologia, Facultat de Farmàcia, Barcelona, Spain. Expert in food science/nutrition.
- Professor Dr Clyde Williams, Loughborough University, Department of Physical Education and Sports Science, England. Expert in exercise physiology/energy metabolism.
- Dr Ron J. Maughan, University Medical School, Department of Environmental and Occupational Medicine, Aberdeen, Scotland. Expert in exercise physiology/energy metabolism/ oral rehydration.

- Professor Dr Sigmund B. Strömme, Norwegian College of Physical Education and Sport, Oslo, Norway. Expert in exercise physiology/sport nutrition.
- Univ.-Doz. Prim. Dr Peter Baumgartl, A. O. Bezirkskrankenhaus, Abteilung für Herz-, Kreislauf- und Sportmedizin, St Johann i.T., Austria. Expert in exercise physiology/energy metabolism.
- Professor Dr Michael Hamm, Fachhochschule Hamburg, Fachbereich Ernährung und Hauswirtschaft, Hamburg, Germany. Expert in human nutrition/sport nutrition.
- Dr Klaus-Jürgen Moch, Institut für Ernährungswissenschaft, Giessen, Germany. Expert in human nutrition/sport nutrition.
- Professor Dr Michel Rieu, Laboratoire de physiologie des adaptations, CHU Cochin, Paris, France. Expert in exercise physiology/energy metabolism.
- Dr Charles-Yannick Guezennec, Division Physiologie-Métabolique, CERMA, Centre d'essais en vol, Paris, France. Expert in exercise physiology/energy metabolism.
- Dr Nancy J. Rehrer, Vrij Universiteit, Dept Sports Medicine, Brussels, Belgium. Expert in nutrition/sport nutrition.

1
Introduction

One of the most important nutritional aspects concerning athletes, recognized since the first competitions in ancient Greece, is the increased need for energy. Athletes involved in heavy physical activity need more food than more sedentary, less active people.

The energy expenditure of a sedentary adult female/male amounts to approximately 2000–2800 kcal per day. Physical activity by means of training or competition will increase the daily energy expenditure by 500 to >1000 kcal per hour, depending on physical fitness, duration, type and intensity of sport. For this reason athletes must adapt their energy intake by increased food consumption, according to the level of daily energy expenditure, in order to meet energy needs. This increased food intake should be well balanced with respect to the macronutrients (carbohydrate, fat and protein) and micronutrients (vitamins, minerals and trace elements). However, this is not always easy. Many athletic events are characterized by extremely high exercise intensities. As a result, energy expenditure over a short period of time may be extremely high. Running a marathon, for example, costs about 2500–3000 kcal (137). Depending on the time needed to finish, this may induce an energy expenditure of approximately 750 kcal/h in a recreational athlete and 1500 kcal/h in the elite athlete who finishes in approximately 2 to 2.5 hours. A professional cycling race, such as the "Tour de France", costs about 6500 kcal per day, a

Figure 1 In professional cycling energy expenditure may exceed 9000 kcal (37 MJ) per day when cycling in the Alps

figure which will be increased to approximately 9000 kcal/day when cycling over a mountain pass (165).

Compensating for such high energy expenditure by ingesting normal solid meals will pose a problem to any athlete involved in such competitions, since digestion and absorption processes will be impaired during intensive physical activity. These problems are not restricted to competition days. During intensive training days, energy expenditure is also high (24). In such circumstances athletes tend to ingest a large number of "in between meals", often composed of energy rich snacks, which, however, are often low in protein and micronutrients. As such, the diet often becomes imbalanced. Especially adapted nutrition/foods/fluids which are easily digestible and rapidly absorbable may solve this problem (23, 30).

During endurance sports activity the body will also use its own energy stores (fat stored as adipose tissue and carbohydrate (CHO) stored as glycogen in liver and muscle). In addition,

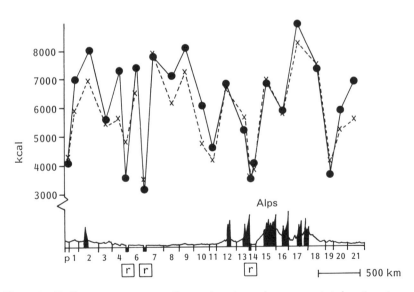

Figure 2 *Daily energy expenditure (•—•) and energy intake (∗---∗) as measured in a cyclist during the Tour de France. Interesting is the extremely high energy expenditure as well as the ability to match energy balance with the use of liquid nutrition in addition to the normal meals p = stage; r = rest day. Reproduced with permission from Saris* et al. *(165)*

small amounts of functional proteins (in the liver, gastro-intestinal tract and muscle) will be broken down due to mechanical and metabolic stresses. These losses have to be compensated by supply of the necessary nutrients. At the same time heat will be produced which to a large extent will be eliminated by production and evaporation of sweat. As a result, fluid and electrolytes will be lost.

Large sweat losses may pose a risk to health by inducing severe dehydration, impaired blood circulation and heat transfer, leading to heat exhaustion and collapse (129, 143, 166, 167, 168). Insufficient replacement of CHO may lead to hypoglycemia, central fatigue and exhaustion (16, 39, 43, 44, 113, 136, 137, 172, 173, 174, 189). Inadequate protein intake induces protein loss, especially of muscle, and consequently a negative nitrogen balance and reduced performance (106, 107, 125).

These observations show that the increased need for specific nutrients should be met according to the level of daily physical activity and exercise. These requirements depend on the type, intensity and duration of the physical effort. Depending on these factors, specific nutritional measures and dietary interventions can be taken, particularly in the phases of preparation, competition and recuperation.

These problems and related nutritional measures concern not only highly trained top athletes but also less trained sports people, especially since the latter, at equal workloads, are prone to more stress, sweat more profusely, use more CHO as fuel for muscle work, utilize/break down more protein and recover more slowly from exhaustion. Highly trained individuals will work more economically and spend less energy to attain certain mechanical work output than untrained people.

Any individual trying to achieve a personal best and exercising at the upper level of functional capacities induces maximal metabolic demands, whether they are an olympic athlete or a recreational athlete trying to complete a marathon.

Well trained athletes have developed a larger metabolic capacity. Therefore, they are able to run faster, and to recover

Figure 3 *Recreational runners performing long distance competitions and trying to set a new personal best perform top sport*

Figure 4 *Individuals engaged in sport and ingesting low energetic diets are prone to marginal or insufficient intake of key nutrients*

more quickly. However, when exercising at maximal capacity, the trained athlete will also become energy depleted, dehydrated and exhausted. Therefore, training and exercise guidelines including nutritional measures and the use of specific foods/meals designed for both categories of athletes should basically not differ.

Foods and meals to be ingested shortly before and during exercise, or during a small break between exercise periods, should be adapted to specific ingestion and assimilation conditions, which depend on the nature and circumstances of the sport practised.

Some groups of athletes compete in sport events where a low body weight is necessary to perform well or to compete in a certain weight category. These athletes are on the one hand training frequently and intensively, but on the other hand they have to maintain a low body weight. The low energy intakes may in these situations lead to a low intake of essential nutrients such as protein, iron, calcium, zinc, magnesium and vitamins; the required CHO intake to balance the CHO used in training may also be marginal.

This aspect should receive special attention since many of these athletes are young and still in a period of growth and development (163).

Depending on the type of sport and training it is possible to categorize athletes at risk for marginal nutrient intake. These athletes and those who combine heavy training with weight reduction programs should be advised most intensively and may benefit from nutritional supplementation (Table 1).

The ability to take safe nutritional measures and to achieve adequate supplementation depends on the availability of standards and guidelines. Safe selection of supplements should be made possible by the establishment of food regulations for this category of products.

Fitness and health oriented people should on the one hand be informed by health/sports medicine organizations and sports federations about the factors affecting food selection, food intake, nutrient utilization and the possible needs for nutritional

Table 1 *High risk sports for marginal nutrition*

Criteria	Sports discipline
Low weight—chronically low energy intakes to achieve low body fat	Gymnastics, jockeys Ballet, dancing, rhythmic gymnastics, ice dancing, aerobics
Competition weight—drastic weight loss regimens to achieve desired weight category	Weight class sports (e.g. judo, boxing, wrestling, rowing, ski jumping)
Low fat—drastic weight loss to achieve lowest possible body fat	Body building
Vegetarian athletes	Especially in endurance events

supplementation (180); on the other hand they should be able to make a safe choice from available food items, including food products designed for athletes. In particular, the nutritional education of athletes and coaches warrants attention. Several studies have shown that nutritional knowledge is marginal, despite the fact that awareness about the importance of nutrition is growing and articles about nutrition are regularly published in athletic journals (46, 149, 198).

Food regulations exist for foods covering special needs in special circumstances, e.g. dietetic food products. However, no such regulation is available for sport nutrition products/ supplements.

The aim of a food product regulation is to lay down scientifically acceptable conditions, such as to ensure that food products are of an acceptable quality standard, i.e. they take into account the specific requirements with respect to nutrients, nutrient mixture compositions and digestibility characteristics. This should also apply to food products which are labeled as sports food. One general problem in this respect is that the current recommended dietary intakes, which differentiate for age, sex and daily activity level, do not seem to be appropriate, for athletes for a number of nutrients such as protein, iron, calcium, zinc and copper.

The aim of this book is to describe nutritional aspects of sport, in particular those related to the macro- and micronutrients

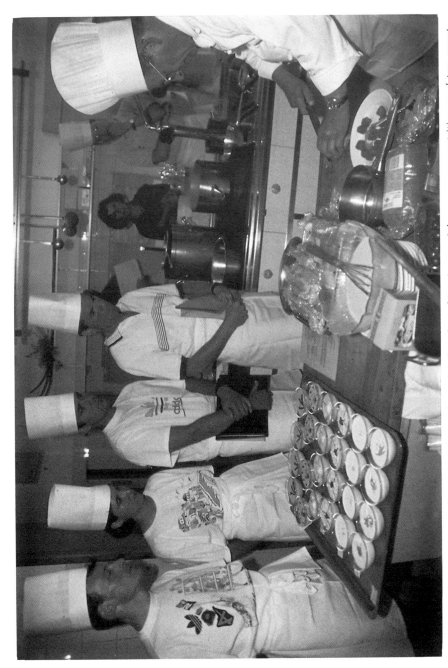

Figure 5 *Swiss national coaches learn about food and cooking in special courses on nutrition and top sport*

which make up the daily nutrition of individuals involved in heavy physical work or exercise. These aspects may form a basis for education in the field of sport and nutrition and may at the same time present the foundation for a possible sports food regulation. Recently major scientific reviews on these topics have been published. One complete review was the output of an international scientific consensus meeting on nutrition and sport (March 1991, Lausanne, Switzerland, proceedings published in a special issue of the *Journal of Sport Sciences*, Volume 9, summer 1991). The development in knowledge, based on sound research over the last 10 years, has been tremendous. The references are listed at the end of the book and supply the interested reader with full details on the topics described.

2
Sport Nutritional Aspects of the Macronutrients

CARBOHYDRATE (CHO)

CHO is the most important fuel for high intensity physical performance. To demonstrate the importance of CHO for performance and recovery we shall briefly describe how CHO makes up a part of the energy reserve and how CHO metabolism and reserves are influenced by exercise (see also Chapter 7).

Carbohydrate reserves

In the body CHO is stored as long chains of glucose units in the liver and in the muscles, called glycogen. This form of storage is comparable to that of starch present in potatoes, banana and other plant foods.

Liver glycogen

The glycogen content of the liver reaches amounts of approximately 100 g. This quantity will vary periodically depending on the amount of glycogen broken down for the supply of blood glucose and the amount of glucose supplied to the liver after

Figure 6 *Different types of starchy food, which are optimal carbohydrate/energy suppliers for intensive sports*

food intake. A constant blood glucose level, within a narrow physiological range, is important since this is the primary energy source for the nervous system. Liver glycogen reserves increase after meals but diminish in between, especially during the night, as the liver constantly delivers glucose into the bloodstream to maintain a normal blood glucose level (89, 90, 91, 135, 158).

Influence of exercise

During physical exercise a number of metabolic and hormonal stimuli will lead to an increased uptake of blood glucose by the working muscles to deliver energy for contractions. To avoid the blood glucose level falling to a too low level, the liver will at the same time be stimulated to supply glucose, mainly from the liver glycogen pool and to a small degree from gluconeogenesis, to the blood stream (1, 90, 135, 158).

Thus glycogen availability in the liver is the key factor for maintenance of a normal blood glucose level during exercise. As soon as the liver glycogen store is emptied and blood glucose utilization by active tissues remains high, blood glucose will fall to hypoglycemic levels. This stressful condition will induce maximal fat mobilization and also protein breakdown and utilization. Glucose uptake by the muscles will drop to marginal levels and the working muscles will then totally depend on local CHO supply systems or indirect supply induced by CHO feedings. Central as well as local fatigue may then occur. This phenomenon has been well described both in sports practice and in scientific studies (39, 44, 63, 65, 91, 173, 174).

Muscle glycogen

Glycogen stored in total muscle tissue amounts to approximately 300 g in sedentary people and may be increased to >500 g in highly trained individuals by the combination of exercise and a CHO rich diet (16, 90, 174). The total intramuscular stored CHO may thus range in energetic equivalent from 1200 to 2000 kcal.

Influence of exercise

The rate at which muscle glycogen is used for the production of energy, needed for muscle contraction, depends on training status on the one hand and duration and intensity of exercise on the other.

Research has shown that, apart from the very small energy rich pool which is immediately present as energy rich phosphates (adenosine triphosphate and creatine phosphate) and will deliver energy for a period of up to 15 seconds, the majority of the energy liberated during muscle work is derived from two main fuel pools—CHO and fat (15, 16, 19, 90, 136, 137, 173).

Use of these two pools will never be exclusive. However, depending on exercise intensity, one of the fuels may become the major energy deliverer. For example, at rest practically all the energy needed for the resting metabolism is derived from fat, with exception of the central nervous system and the red blood cells, which rely on blood glucose. The possible energy supply ratio in this situation may be in the order of 90% fat to 10% CHO.

During a situation of greater activity, i.e. physical work, or a moderately intensive sport activity, the body under the influence of metabolic, hormonal and nervous mechanisms, will additionally mobilize glucose from the liver and from the muscle glycogen pool, to deliver energy (133). At the same time the mobilization of fatty acids will increase until a metabolic steady state has been achieved (after approximately 20 minutes). The energy supply ratio of fat to CHO might then be 50% : 50%.

At higher intensities the body will start to use more and more CHO. This means that during highly intensive sports activities CHO will become the most important fuel (39, 44, 79, 113, 174). The ratio of fat to CHO may then be 10% : 90%.

The reason for this shift to the dominant use of CHO is that the maximal amount of energy which can be produced from CHO is per unit of time higher than that of fat. In addition, the amount of oxygen needed for energy production from CHO is about 10% lower than that of fat (119).

These factors enable athletes to work at a higher intensity when using CHO as the main energy source. Indeed, research has shown that exercising individuals whose muscle and liver glycogen levels have been emptied, either by exercise or by a combination of exercise and low CHO intake, can only work at about 50% of their maximal working capacity (39, 113, 137).

Alternatively, when CHO stores in muscle and liver are increased by diet manipulation, athletes are able to perform longer at a high exercise intensity. These examples show that the size of the glycogen pool is one of the limiting factors in endurance performance. As soon as specific muscles or muscle fibers have depleted their glycogen stores they will be impaired in their ability to perform repeated high intensity contractions (16, 39, 90, 113).

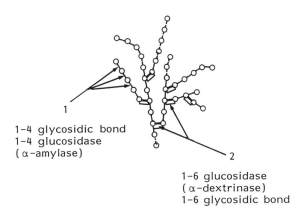

1
1-4 glycosidic bond
1-4 glucosidase
(α-amylase)

2
1-6 glucosidase
(α-dextrinase)
1-6 glycosidic bond

Figure 7 *Starch is built up from glucose molecules which are connected by 1-4 and 1-6 bonds. Both bonds are split by specific enzymes during the process of digestion. Glucose, the end-product of starch digestion, is absorbed by the gut and delivered to the blood. After uptake by the muscle and the liver, glucose can be restored in the form of glycogen, which has a structure comparable to that of starch*

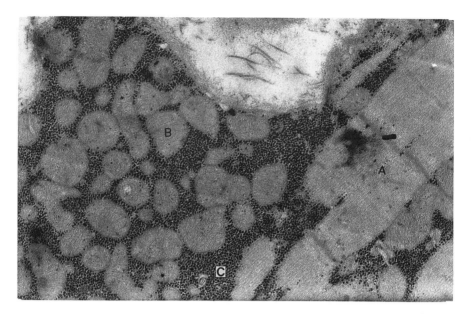

Figure 8 *Muscle glycogen is stored as "starch" granules (C) within the muscle fibers (A) between the mitochondria (B). The amount of glycogen stored can be measured by biochemical analysis of a muscle biopsy sample*

Time course of glycogen depletion

Four important factors determine the speed and the extent to which CHO stores will be emptied.

1. Exercise intensity.
2. Exercise duration.
3. Training status.
4. CHO ingestion.

As explained above the use of glycogen depends primarily on exercise intensity and duration. At low to moderate intensities, fat will additionally serve as a potential energy source. Thus CHO reserves may be emptied slowly at low intensities, i.e. in 4 hours when exercise intensity approximates 55% of VO_2max, in about $1\frac{1}{2}$ hours at 65% VO_2max, and in less than $1\frac{1}{2}$ hours at higher intensities such as during highly intensive training sessions (interval, tempo) or ball game sports (soccer, ice hockey, rugby etc.) (44). (VO_2max means maximal oxygen uptake. Oxygen uptake increases with increasing exercise intensity until a maximum is achieved. The exercise intensity at this point is determined as 100%).

Training status: The time course will further be influenced by training status since highly trained individuals have developed an enhanced capacity to use fat as energy source compared to less trained individuals. Thus, when working at the same absolute exercise intensity (e.g. running at a speed of 15 km/h), trained individuals will use less CHO and more fat for muscle contraction (19, 71). Under competition circumstances, however,

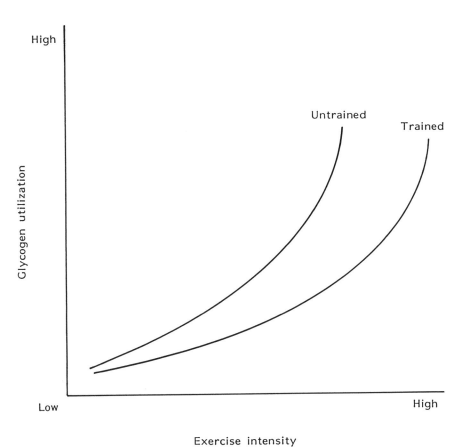

Figure 9 *Glycogen depletion rate depends on exercise intensity. Trained subjects use less glycogen and more fat at submaximal exercise intensities*

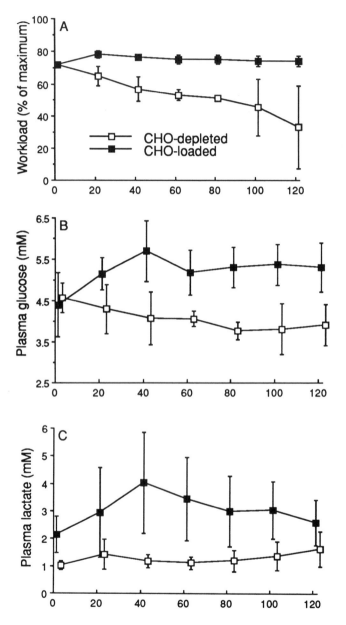

Figure 10 *Metabolism in a glycogen depleted state. Because of lack of glycogen in the liver, blood glucose falls, lactic acid production is decreased and fat metabolism is increased to compensate for energy deficits. The*

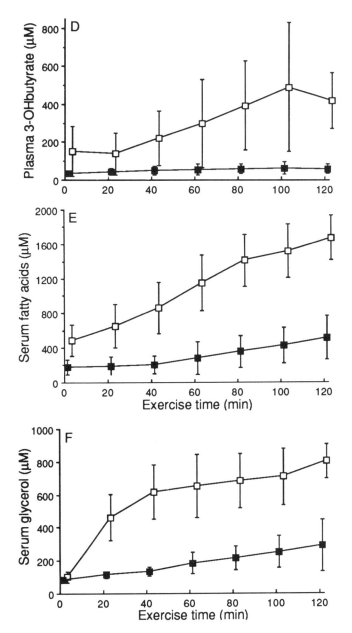

consequence is that the performance level will drop to approximately 50% of maximal capacity. Reproduced from Wagenmakers (189), with permission from the American Physiological Society

the latter will not be the case since any individual then works at maximal capacity. Thus, when working at the same relative intensity (e.g. 80% of maximal working capacity) CHO and fat utilization may not differ between trained and untrained subjects.

Carbohydrate supplementation during exercise

The rate of utilization of CHO stored in the body can further be reduced by supplying the blood and muscle tissue from an alternative source, i.e. oral CHO. When CHO is ingested, digested and absorbed, it will enter the portal bloodstream, pass to the liver and finally enter into the circulation. Blood glucose will be increased as a result of oral CHO intake. On one hand this reduces the need to break down liver glycogen for maintenance of blood glucose. On the other hand this will enhance glucose supply to the muscle and glucose uptake by the muscle. Indeed, a large body of scientific evidence shows that oral CHO intake reduces liver glucose output but increases blood glucose and consequently glucose uptake by the muscle.

In addition, oral CHO is oxidized during exercise. Theoretically the latter will reduce the rate of muscle glycogen breakdown for energy production, reduce the catabolism of protein and delay fatigue/improve performance (49, 75, 124, 148).

Thus, in studies where CHO was ingested during exercise, total CHO utilization was found to be the same as in non-CHO ingesting control groups. Since oral CHO was shown to be oxidized the conclusion is that glycogen must be spared. However, CHO ingestion has not been found to reduce the rate of muscle glycogen breakdown in active muscle groups (35). Any sparing of endogenous CHO is thus most probably in the liver or in muscle groups not actively involved in the exercise.

There is one prerequisite to achieving sparing or build-up of endogenous CHO pools during exercise: ingested CHO should be easily digestible and rapidly absorbed.

For exercise lasting longer than 45 minutes it is generally recommended that at least 20 g, but optimally up to 80 g, be consumed for every following hour of exercise (39, 44). Such amounts have been shown not to delay gastric emptying to a physiologically important extent and to increase water absorption in the intestine. This aspect is of particular importance in endurance events in the heat, where fluid availability may be the first priority (see Chapter 3).

During endurance exercise, low amounts of CHO may only be sufficient to maintain normal blood glucose levels, whereas higher amounts may more effectively decrease the use of endogenous CHO stores and delay fatigue. The CHO sources used should be rapidly digestible and absorbable. Most efficient are (soluble) CHO sources which can be ingested with fluid. The gastric emptying rate should be relatively fast and the physical form of the CHO should allow rapid digestion/ enzymatic hydrolysis. This is not the case with all CHO sources. For example, the dietary fiber in which some CHO sources are "packed" may form a physical barrier to digestive enzymes (47) and may also reduce gastric emptying rate. In contrast to normal daily meals—which should primarily contain slowly digestible CHO sources, rich in dietary fiber and having a low glycemic index, such as whole grain products— foods taken shortly before and during exercise should be low in dietary fiber and have a high glycemia index (30, 44).

The reason for this paradox is that dietary fiber reduces gastric emptying and decreases the extent to which enzymes can immediately be effective in hydrolysis. It also increases gastro-intestinal bulk, enhances transit in the gut and may be subject to bacterial fermentation causing gas production. Softening of the intestinal contents by fiber and the related improved intestinal transit are desirable in sedentary individuals but pose a problem during intensive exercise. These factors together may explain why athletes ingesting slowly digestible whole grain foods, prior to and during exercise, experience more gastrointestinal problems (30, 156).

When dietary fiber is excluded from the CHO source, starch is shown to be as effective in energy supply as free glucose (75).

Other sources of complex CHO, such as rice, spaghetti and potato, are of particular interest for daily CHO intake, between

Figure 11 The difference between "raw" carbohydrate sources and refined sources is the dietary fiber content. Dietary fiber reduces gastric emptying rate, slows down digestion and absorption and enhances the amount of intestinal bulk, which promotes normal transit

Figure 12   *Carbohydrate loading*

sport sessions, but are shown to be oxidized more slowly during exercise than soluble CHO sources (76). During periods of non-intensive exercise, however, such as mountain walking, these CHO sources can be consumed satisfactorily.

Optimal CHO sources for high intensity endurance events are processed (pre-digested) CHO low in dietary fiber:

● Monosaccharides (glucose).
● Disaccharides (sucrose, maltose).
● Polymers (maltodextrins, malt extract).
● Starches (soluble starch).

These types of CHO have the additional benefit of being easily dissolved in fluids, a very important aspect, since CHO and fluid needs (see chapter 3) are both determined by exercise intensity and duration. The types of CHO listed above have been shown to be about equally effective in increasing blood glucose levels and oxidation rates during exercise and in improving performance (43, 44, 82). Also, effects on blood insulin levels during exercise do not appear to be different (44). Some early studies appeared to show that intake of 50–75 g of rapidly absorbable CHO prior to exercise induces a rapid rise in blood glucose and insulin, which, when exercise is started, leads to a rebound hypoglycemia and decreased performance. However, these studies were done after an overnight fast and CHO was ingested in the resting state, 45–60 minutes prior to exercise. These conditions are not comparable to those of the endurance athlete, who will eat a pre-game breakfast and perform a warming-up. Studies done under practice conditions did not reveal any rebound hypoglycemia (26, 27). Meanwhile a large number of studies have shown that pre-game CHO intake is beneficial in delaying fatigue (for review see 44, 45).

One exception may be pure fructose, which on the one hand has been shown to maintain euglycemia and not to influence insulin secretion, resulting in a less strong inhibition of free fatty acid mobilization than glucose, and on the other hand is slowly absorbed. Some authors have reported intestinal upset when >30 g/l is ingested at rest and during exercise (128). However, in one study this effect, with ingestion of up to 1 g/kg body weight during exercise (49), was not found.

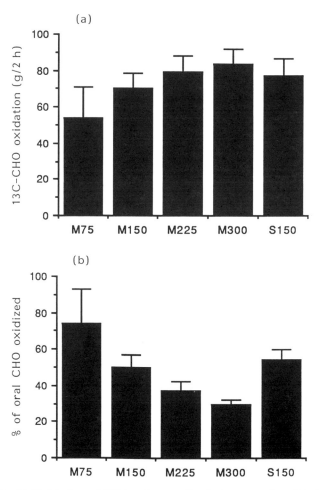

Figure 13 *(a) Oxidation of CHO taken orally during exercise. (b) Oxidation of oral CHO as a percentage of the given dose (grams). M, maldex-20, S, sucrose. Oral carbohydrate is oxidized during exercise, and thus contributes to energy production. Increasing CHO intake to about 100 g/hour increases its oxidation. Higher CHO intakes have no effect, most probably because of a delayed gastric emptying. This is clearly represented by the percentage of CHO oxidized. Interestingly glucose polymers (maltodextrins) are as well oxidized as sucrose. From data of Wagenmakers et al. (213)*

Additionally, oxidation rates of fructose in energy delivery processes have been shown to be lower, most probably because of a higher affinity of the enzyme hexokinase in muscle for glucose, compared to fructose. This makes pure fructose as a high energy source during exercise less attractive (24, 44, 45,

49, 112, 128). In low concentrations, <35 g/l, or in combination with glucose supplying CHO sources (e.g. glucose, sucrose, maltose, maltodextrins, soluble starch), fructose may not induce gastrointestinal distress and will supply an extra energy source for the liver. It has been shown that when fructose is taken in equal amounts with glucose or is taken as fructose, its absorption is enhanced and is dose dependent (160).

With respect to palatability and gastrointestinal comfort, starch hydrolysates (maltodextrin/glucose polymers) and soluble starch may have the benefit of being less sweet than the mono- and disaccharides. They also have less effect on fluid osmolality and have been shown to increase quantitative glucose absorption, which is of advantage at higher CHO concentrations (155). At higher concentrations (100–200 g/l) drinks would be strongly hypertonic with dissolved mono-/disaccharides but not with maltodextrins/polymers or starch.

Carbohydrate intake at rest

After exercise the endogenous CHO pools should be replenished. Depending on the time available for total recovery, i.e. the time elapsing until the next sport activity, there may or may not be a need for fast replenishment.

Glycogen synthesis has been shown to be most rapid during the first hours after exercise. Thereafter the synthesis rate will gradually decline (44, 45, 90). Glycogen synthesis itself is only possible if the needed building substances, i.e. glucose molecules, are supplied. Net glycogen synthesis rate, there-fore, depends on synthesis capacity and quantitative glucose supply (39, 44, 45). The latter depends largely on the type of food ingested, i.e. the rate of digestion and absorption. The CHO source may also be important. Glucose favours muscle glycogen recovery, whereas fructose is primarily taken up by the liver, thus favouring liver glycogen recovery (17, 91).

If the next activity takes place after one or two days, the athlete should recover by taking normal meals with a high CHO content (55–65 en%) (en% = percentage of total daily energy

Figure 14 *Six day Tour de France simulation experiment in a respiration chamber at the University of Limburg, Maastricht, The Netherlands. Indirect calorimetry allows continuous measurement of energy expenditure*

intake) and composed of low glycemic index foods such as whole grains, fruits, vegetables etc. A slow digestion and absorption rate is favorable in this condition. Under these circumstances 400–600 g of CHO/day should be sufficient to recover glycogen depots at energy expenditures up to 4000 kcal (39). However, if daily energy expenditures are extremely high, such as during multiday cycling competitions, and exercise intensity is high, the CHO need may then be >12 g/kg body weight/day. CHO intake with normal meals alone results in too much bulk and causes gastrointestinal distress. Therefore, athletes who ingest normal solid food will undereat on days of prolonged intense exercise, which will result in a negative energy balance and an insufficient CHO intake. These high needs can then only be covered with additional ingestion of CHO dense foods/solutions (25, 101, 165). Additionally, if the time for recovery is limited, i.e. the second training session or competition will follow on the same day, then the meal in between should be composed of foods which are rapidly digested and absorbed, i.e. a high glycemic index. (Foods that lead to a slow increase in blood glucose have a low glycemic index. Those that induce a rapid rise in blood glucose have a high glycemic index.) Processed, cooked and blenderized potato, rice, noodles or corn starches belong to this category. CHO solutions can be taken during exercise and in any situation in which CHO intake from normal food cannot take place or is insufficient and will help to enhance glycogen recovery in the first few hours after exercise (44, 45, 53, 98).

FAT

Fat forms the second main energy source for the exercising individual. The importance of fat as an energy source depends on the degree of exercise stress as well as on the availability of CHO. We shall briefly describe how fat is part of the total energy reserve in the body and how fat utilization and the fat depots are influenced by exercise. For a schematic presentation of fat metabolism see chapter 7.

Fat reserves

In non-trained healthy people the body fat content may be 20–35% for females and 10–20% for males. Fat is stored in the body as triglycerides in fat cells (adipocytes) which make up the adipose tissue. Additionally, a small fraction of triglycerides is stored within the muscle cells and circulates in the blood bound to albumin.

Adipose tissue

The major part of adipose tissue can be found under the skin, the so-called subcutaneous fat tissue. In addition, fat is predominantly stored around the abdominal organs.

Depending on long-term nutritional conditions, this fat storage can either become minimal in case of long-term negative energy balance, such as during periods of anorexia and fasting, or become very large in case of long-term positive energy balance, such as during chronic overeating.

In highly trained people the total fat stored in adipose tissue— 5–15% in males and 10–25% in females—is less (203) than that in sedentary subjects—20–35% for females and 10–20% for males. Nevertheless this amount of fat has a very large energy potential (approximately 7000 kcal/kg adipose tissue), which makes adipose tissue the most important energy store in any case of prolonged energy deficit in which the body CHO stores become progressively depleted and fat becomes the main energy fuel. This may be during chronic food deprivation, but also during shorter periods of high energy expenditure resulting in negative energy (CHO) balance (19, 134, 136, 138). However, some CHO is always needed to provide the necessary intermediates for the citric acid cycle (substances needed to keep aerobic energy production continuing). For this reason the body will start to produce glucose from other substrates (a process called gluconeogenesis) in any situation where a lack of blood glucose exists (135, 138).

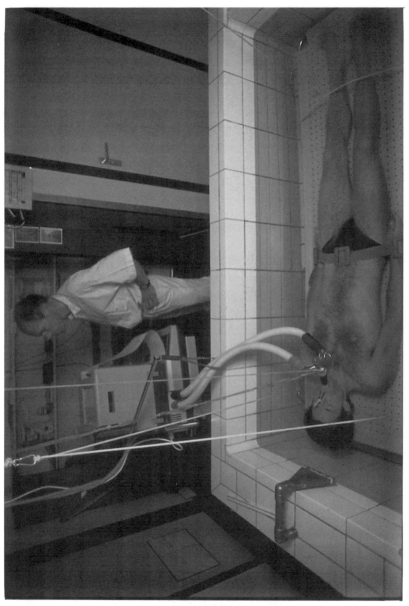

Figure 15   Determination of lean body mass and body fat content by hydrostatic weighing. Lung volume is determined and corrected for by a helium dilution technique (photograph R.u.L.)

Influence of exercise

During physical exercise a number of nervous, metabolic and hormonal stimuli will lead to an increased rate of fat utilization on the one hand and fat mobilization on the other.

Fat in the form of free fatty acids (FFAs) will be increasingly oxidized within the mitochondria of the muscle cells. As a result the concentration of FFAs within the muscle cell will fall, which will stimulate uptake of FFAs from the blood. Increased blood flow to the muscle is the first step in delivery of more FFAs to the muscle cells. This process of FFA transport, uptake and mobilization is stimulated by action of so-called stress hormones (adrenalin and nor-adrenalin) which will increase while exercising and stimulate lipolysis, by a reduction of circulating insulin and increased activity of the central nervous system (19, 134, 138).

The steps to realize an enhanced fat oxidation are numerous and complex. This is the major reason why achieving a steady state adaptation will take about 20 minutes.

For this reason CHO utilization has to compensate for any shortage of energy in this initial adaptation phase, for as long as the utilization capacity and energy production from fat is not at a maximal level (1, 136). Energy production from CHO is also "faster" than from fat (119).

FFAs from the adipose tissue store will be available for a very long time, once increased fat mobilization transport and uptake—resulting in a metabolic steady state—has been achieved. If fat was the only substrate, this would theoretically enable individuals to run continuously at marathon speed for >70 hours, equivalent to an energy expenditure of >70 000 kcal (136).

However, this would only be possible if fat could deliver an adequate amount of energy and if pain in the musculoskeletal system did not occur. At competitive speeds, CHO availability will be one of the factors limiting performance times, since fat alone forms an inadequate energy source when exercising at high intensity (136, 137).

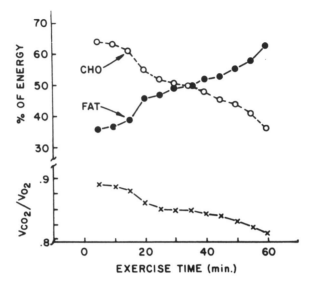

Figure 16 *Fat mobilization, transport and utilization are relatively slow processes. At the onset of physical exercise most energy comes from CHO metabolism. After approximately 20 minutes fat metabolism proceeds at full speed and CHO utilization will be reduced. Reproduced from D. Costill, J Appl Physiol 1979; 47: 787–791, with permission from Association Colloques Physiologie, St Etiennes, France*

Regular endurance training increases the capacity of the skeletal muscle to use fat as energy source. An increase in the utilization of fat as an energy source during endurance activities will, when working at a fixed exercise intensity, enable the athlete to reduce the use of CHO. This in turn will spare endogenous CHO, and may delay fatigue.

In addition, fat cells will increase their sensitivity to stimuli for FFA mobilization, thereby improving the speed of adaptation to increased needs when exercising (19).

During maximal exercise intensity, however, endogenous CHO utilization seems to proceed at full speed and enhancement of blood FFAs does not automatically lead to a reduction of muscle and liver glycogen utilization (7, 162).

Muscle fat

Within muscle, fat is stored as triglycerides in the form of small fat droplets, located near the mitochondria. However, this part of the fat storage represents only a fraction of the total storage. Although endurance trained individuals possess less adipose tissue than sedentary people, their muscle fat content is found to be larger (25). This leads to the question "why"? One possible reason may be that endurance exercise leads to a partial depletion of intramuscular fat. An increase of this fat store would thus mean an increase in substrate availability. As

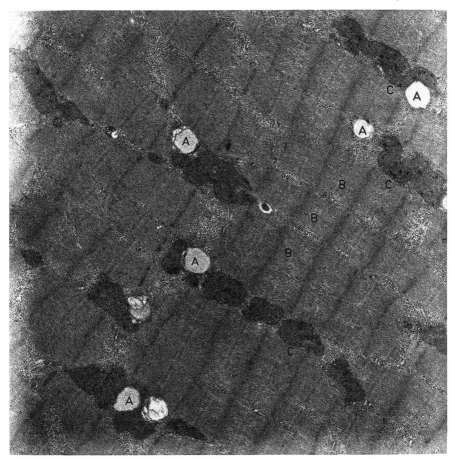

Figure 17 *Muscle tissue contains fat in the form of small fat droplets (A) stored within muscle fibers (B), near mitochondrial networks (C). Endurance trained athletes possess more intramuscular fat*

such, this would be a normal physiological adaptation. In proportion to total body fat, intramuscular fat content is very small.

Influence of exercise

Scientific evidence shows that the influence of exercise on muscle triglycerides may be the same as on adipocytes. Decreasing intracellular FFA concentrations as well as neural and hormonal stimuli enhance lipolysis leading to the liberation of FFA which will be taken up by the mitochondria for oxidative energy production.

As a result, intramuscular fat stores have been shown to be decreased after endurance exercise (19, 25).

Fat intake

Sedentary people living in industrialized countries consume diets which contain 35–45% of total energy content as fat (19, 131). These figures are relatively high, seen in the light of recommendations that daily food should be rich in CHO (>50 en%) (en% = percentage of daily energy intake). Athletes are generally advised to reduce fat intake to approximately 20–30 en%, thereby enhancing CHO intake to 60–70 en% (19, 39, 44, 45, 90, 173). This reduced fat intake should to a large extent be realized by consumption of lean meat, low fat processed foods and limiting consumption of fried and other fatty foods. Saturated fatty acid intake should be limited to less than 10 en%, mainly by use of plant oils for meal preparation. With improved food quality and increased total energy consumption, the low figure of 20–30 en% fat intake in athletes will lead to a more than sufficient supply of the essential fatty acids needed for normal biological functions (at least >1 en%, preferably about 7 en% (131)).

Fat supplementation

For individuals involved in heavy physical work or sports activities there seems to be no rationale for fat supplementation during exercise. The fat store in the body is large enough to compensate for any need. Moreover, lipolytic stimulation during exercise will enhance blood FFA levels to such an extent that a maximal rate of FFA uptake by the muscle cells and mitochondria is achieved. Oral supply of fat may additionally increase blood FFA levels but not uptake in active muscle cells and oxidation rates (19), and may thus not be of benefit in reducing muscle and liver glycogen utilization (7, 162).

In contrast, therefore, the task is to reduce total fat intake, making a larger CHO consumption possible, while at the same time maintaining the energy balance. Foods/meals especially designed (they should be light digestible and rapidly absorbable) for ultraendurance events such as triathlon, high altitude climbing, multiday endurance events etc., and which replace normal CHO rich solid meals, may be mixtures of easily digestible CHO, fat and protein. Such foods, which replace the normal meals under exercise conditions, should contain $\leqslant 30$ en% fat.

Polyunsaturated fatty acids are known to influence the structure of the cell membrane, especially of red blood cells. Increased intake of omega-3 fatty acids, by means of supplementation, resulting in increased red blood cell plasticity, maximal oxygen consumption and better blood oxygen levels in subjects exercising at altitude, has been reported (77).

Medium chain triglycerides (MCT), which are taken up rapidly in the intestine (as CHO) and are easily transported into the mitochondria (22), may be an interesting component for competition meals of endurance athletes. Oral MCT is found to be readily oxidized during exercise (7, 48, 111) and could thus serve as substrate during ultraendurance activities which proceed at lower exercise intensity. Ingestion of MCT prior to exercise has not been shown to improve performance (7, 92, 162). Thus, research on the influence of MCT fat on gastric emptying, absorption and oxidation during exercise, when

Figure 18   *Fish is an excellent source of polyunsaturated fatty acids. Omega-3-fatty acids are important for membrane plasticity and stress tolerance of red blood cells*

ingested together with CHO, is required to draw definite conclusions about possible benefits.

PROTEIN

Protein forms an important basis for growth and development of organs and tissues. Growth requires amino acids as building substances and an insufficient supply of nitrogen in general or of essential amino acids (those which cannot be synthesized by the human body) in particular is known to be associated with impaired growth, especially of muscle tissue, as well as impaired health. Here we shall briefly describe how protein forms a part of important biological functions and how these are influenced by exercise. For a schematic presentation of protein metabolism see Chapter 7.

Protein reserves

The human body has no protein reserve/store comparable to its large energy store composed of fat and moderate store of glycogen.

All protein in the body is functional protein, i.e. is either part of tissue structures or part of metabolic systems such as transport systems, hormones etc.

Any protein which is superfluous cannot be stored as protein but will be broken down so that the nitrogen is excreted with urine and the remaining part is either used immediately in energy production or metabolically converted and stored either as glycogen or as fat. The latter will be only a very small fraction.

Nevertheless the body possesses three major functional protein pools of which amino acids may be used under stressful conditions such as food deprivation and energy deficits (25, 63, 64, 71, 106, 107, 125, 138, 139, 157, 187, 188):

1. The plasma proteins and amino acids (AAs).
2. Muscle protein and intracellular AAs.
3. Visceral protein and intracellular AAs.

Plasma proteins/amino acids

Albumin and hemoglobin are two major plasma proteins. Both are involved in transport processes (carriers) and may be reduced as a result of long-term insufficient protein (nitrogen) intake or energy intake, or a combination of both. Other plasma proteins with a rapid turnover, such as pre-albumin and retinol binding protein, respond to short-term changes and are therefore used as markers for nutritional status (161).

Since transport proteins such as hemoglobin form an important part of metabolic chains for energy production, it may be concluded that any reduction will be associated with impaired metabolism and will influence performance capacity. Especially in athletes a reduction in hemoglobin is known to reduce oxygen transport capacity, thereby reducing oxidative energy production capacity and endurance exercise capacity (38, 209).

Plasma AAs form the central pool of metabolically available protein substances. Any protein consumed will, after digestion and absorption, influence the plasma AA pool. Any AAs for synthesis of functional protein will be taken up from this AA pool.

The composition of plasma AAs is kept within a close range. Shortage of non-essential AAs will induce production of these AAs by the body. Shortage of essential AAs on the other hand cannot be compensated by de novo synthesis (synthesis by the body). There are only two ways to compensate for such a shortage: increased consumption of protein containing these essential AAs or breakdown of functional protein within the body, in which these essential AAs are built in. As a result, these AAs are liberated in the plasma pool.

Apart from their need for tissue structures AAs have a large number of key functions in energy and central nervous system metabolism.

AAs play a major role in intermediate metabolism, are precursors for gluconeogenesis, and for hormones and peptides which function as neurotransmitters (64, 71, 80, 99, 106, 107, 125, 149, 157, 187, 188, 204).

Any pronounced change in plasma AA composition can, therefore, initiate changes in protein synthesis rate, alertness, fatigue, mood etc. (206). Any prolonged change may have consequences for health.

Influence of exercise

Exercise is known to be associated with changes in plasma AA composition. It has been shown that branched chain amino acids (BCAAs: leucine, valine, isoleucine) will contribute to energy production during exercise. As a result, their concentration in plasma will fall (1, 106, 107, 187, 188). This has two major consequences:

1. The nitrogen which is split off will lead to the formation of ammonia, a metabolic end-product known to be toxic and associated with fatigue (29, 187, 188).
2. The ratio of BCAAs to other amino acids will change. As a result of this change some AAs which are known to be precursors for hormones and peptides active in the central nervous system, will increasingly pass the blood–brain barrier and increase their concentration in the brain.

As a result neurotransmission and fatigue is thought to be influenced (139). The more pronounced these changes in plasma AAs are, the more pronounced may be the effect on citric acid cycle metabolism and fatigue.

It has been shown that a shortage of CHO (glycogen, blood glucose) dramatically increases the need to use protein (BCAAs) for the production of energy (106, 107, 188).

Recently two major lines of evidence have been reported (187, 188, 189):

1. Depletion of endogenous CHO pools leads to
   a. dramatic changes in intramuscular and plasma AAs;
   b. increases in the activity of enzyme complexes involved in the breakdown and oxidation of BCAAs;
   c. rapidly increasing intramuscular and plasma ammonia levels;

  d. a reduction of the time to exhaustion;
  e. increased nitrogen loss through sweat and urine.
2. Supplementation with CHO, maintaining sufficient endogenous CHO available, minimizes these changes.

Athletic effort always places an energetic stress on the body and will therefore always lead to increased utilization of AAs, some of which are essential. In any endurance event this utilization will be maximized due to depletion of endogenous CHO pools, i.e. will be minimized when sufficient CHO availability is maintained.

Muscle protein

Muscle mass forms the largest protein pool within the body. Apart from the functional aspect that muscle protein has the ability to contract and thereby generate mechanical work, muscle protein has also been thought to be the amino acid

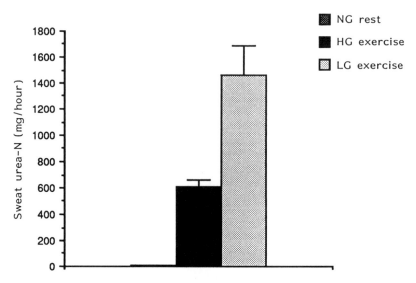

Figure 19 *Influence of CHO availability on nitrogen losses in the form of urea-N in sweat at rest and during exercise. NG, normal glycogen; HG, high glycogen; LG, low glycogen. Reproduced from Lemon and Mullin,* Med Sci Sports Exerc *1981, 3: 141–149, with permission from the American College of Sports Medicine*

Figure 20 *World champion shotputter Werner Günthor. Strength athletes are characterized by a large muscle mass*

supplying pool during starvation conditions (125, 138). Starvation is characterized by a decrease in muscle mass.

Three main goals may be achieved by breakdown of muscle tissue under such circumstances:

1. Liberation of amino acids for use in energy production and maintenance of a normal blood glucose level (gluconeogeneis).
2. Supply of essential amino acids to maintain a normal plasma amino acid composition.
3. Liberation of glutamine to maintain a normal plasma glutamine level, which is assumed to be important for immune status and normal gut function.

Apart from these metabolically important aspects, muscle protein reductions may also occur as a result of changes in the ratio of anabolic and catabolic hormones. Starvation and also physical exhaustion due to energy deficits are known to change the anabolic/catabolic ratio towards catabolism. As a result, de novo synthesis of protein may fall to low levels. Increased degradation and oxidation of protein together with decreased synthesis will then result in a net loss of functional protein, measurable as a negative nitrogen balance.

Influence of exercise

Increased AA oxidation as well as nitrogen losses, induced by exercise have been described in many studies (25, 64, 106, 107, 157, 187, 188) . There has been a debate about whether the AAs oxidized during prolonged exercise stem from the muscle, from the gastrointestinal tract including the liver, or from both. Measurements across particular muscle groups (performed by determination of arteriovenous AA differences) have shown that some amino acids are produced/liberated from the muscle during exercise. However, this may not necessarily reflect net catabolism of muscle tissue since the muscle is also able to synthesize amino acids, e.g. alanine from pyruvate and nitrogen derived from the breakdown of other AAs.

Apart from this, microdamage to muscle fibers due to the influence of mechanical stress may lead to the loss of AAs and proteins such as enzymes. This type of damage known to occur especially in running events and during negative work (eccentric contractions, e.g. downhill walking or running) induces repair and inflammation processes after exercise, which will lead to pain perception, most intensively 2–4 days later (so-called delayed onset of muscle soreness, DOMS) (6). Repair processes require the supply of AAs. However, the damaged muscle cells which are broken down to enable repair will liberate their AAs in the same AA pool as that from which AAs are used for de novo protein synthesis. Mechanically induced catabolism will thus not necessarily lead to net loss of protein/AAs or to an increased requirement.

Visceral protein

After muscles, visceral tissues form the second largest protein pool. Visceral tissues have been observed to contribute significantly to interorgan exchange of AAs during fasting and physical stress induced by illness (125).

Influence of exercise

Exercise may induce an increased contribution of visceral protein to amino acid exchange between organs (157). However, there is some speculation on the quantitative contribution of AAs, derived from the visceral pool, to glucose production in the liver and to nitrogen losses during and after exercise.

Although it was suggested in previous years that the exercise induced nitrogen loss was derived mainly from muscle protein, there are some recent indications that visceral tissues, which undergo a large reduction in blood flow and in some conditions may become ischemic (especially the colon (31)), may make a significant contribution (157). A recent study on the effects of exercise on intestinal protein turnover indicated a reduced protein synthesis and increased protein degradation during exercise (31, 193).

From previous paragraphs it may be concluded that the main reason for net protein (nitrogen) loss as a result of endurance exercise is the utilization of AAs, derived from different pools, in intermediate/energy metabolism. The process is known to be intensified during energetic stress such as a state of high energy needs while being glycogen depleted and leading to negative nitrogen balance (25, 106, 107, 187, 188).

Protein intake

The average recommended daily intake range for protein in European countries is 54–105 g for adult males and 43–81 g for adult females (183). In comparison, the recommended daily allowance (RDA) in the USA amounts to 58 and 50 g respectively or 0.8–0.9 g/kg body weight/day (131). In general the protein intake in healthy people in the Western world expressed as en% of total daily intake, amounts to 10–15 en% (en% = percentage of daily energy intake), resulting in daily intakes of about 50–110 g at energy intakes of 2000–3000 kcal.

This figure does not seem to change very much in populations involved in prolonged heavy exercise. A value of 12 en% was observed during cycling the Tour de France while expending and ingesting 6500 kcal/day over 3 weeks (165).

Therefore, it can be concluded that increased energy intake results in increased protein intake, since the latter seems to be a more or less constant percentage of total energy intake. This relation, however, does not exist in vegetarian athletes, who generally are assumed to have low daily protein intakes, because of a reduced protein density (141) and, in general, a low energy intake (55), or in female athletes who consume only small amounts of food, such as gymnasts and dancers (176) and, surprisingly, long distance runners (151). Interestingly, female athletes consuming low energetic diets and little animal protein are often found to be amenorrheic.

There is a large body of evidence that the protein requirement of endurance athletes ranges from 1.2 to 1.8 g/kg body weight/day (105, 106, 107, 108). There are only limited data on athletes involved in strength sports and possessing a relatively high muscle mass and low fat mass. It is frequently stated that they require more protein than endurance athletes (mainly because of their higher lean body mass) to achieve optimal training status and performance. However, this has often been suggested solely because of their high protein intakes, sometimes >4 g/kg body weight (108).

Only two well-controlled nitrogen balance studies are available on strength athletes. Tarnopolsky *et al.* (182) determined nitrogen balance in six elite bodybuilders, six elite endurance athletes and six non-training control subjects. He observed that endurance athletes require 1.67 times more daily protein than the non-training control subjects. Bodybuilders needed only 1.05 times more protein to maintain nitrogen balance. However, since this study was carried out over only 10 days and the training load was not explicitly described, it is not clear how well these data represent the true situation in periods of fluctuating intensive training. Walberg *et al.* (191) studied weight lifters in a weight loss/training regimen. She observed that 0.8 g protein/kg body weight/day resulted in a negative nitrogen balance whereas two times the RDA, 1.6 g kg/day, resulted in a positive nitrogen balance. These data indicate that protein requirements in strength athletes may be only slightly increased, in contrast to cases with a high energy turnover or consuming low energetic diets, both of which lead to a decrease in blood glycogen stores, which may increase protein utilization. Independent of these nitrogen balance derived data, however, it is generally believed that intakes of 1.5–2.5 g/kg body weight contribute to optimal wellbeing and performance in strength athletes (108).

A recent study on protein overloading in strength athletes, however, demonstrated that when 2 g of protein/kg body weight was supplemented, on top of a normal protein intake with food of 1.3 g/kg, protein turnover more than doubled. That is, both synthesis and breakdown were increased by >100%. As a result nitrogen excretion with urine was also more than doubled. However, during the 4 week training period the strength athletes gained significantly more muscle mass. This suggested that the absolute rate of protein turnover

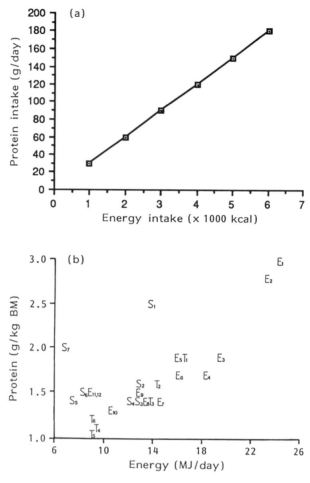

Figure 21 *(a) Protein intake expressed as 12 en% of total daily energy intake. Note that an increase in total energy intake will automatically lead to an increased protein consumption. The latter has been observed to amount about 12 en% in endurance athletes, even during the Tour de France. (b) Protein intake in relation to energy consumption. E = endurance; S = strength; T = team sports. These data show a clear relationship between energy and protein consumption. Athletes who consume less than 1500 kcal may be prone to a marginal or insufficient protein intake. Some protein supplementation to enhance the protein density of the diet may then be advised. Reproduced from Erp-Baart et al., Int J Sports Med 1989, 10, Suppl. 1: S3–S10, with permission from Georg Thieme Verlag, Stuttgart and New York*

in combination with training stimuli may determine in some way the extent to which lean body mass increases (66). Thus, although intakes above a level of 2.0 g/kg body weight seem totally superfluous from a requirement point of view, they may have an anabolic effect. The study cited used ^{15}N glycine as a marker for protein synthesis and degradation rate. Since this label is not universally accepted as appropriate, this study has to be interpreted with care. More research is needed to verify the results.

Protein supplementation

In terms of nutritional requirement, it appears that protein supplementation by increasing daily protein intake to a level higher than 12–15 en% will be superfluous for most athletes.

Since higher energy intake in endurance athletes will result in higher protein intake, the value of protein supplementation for endurance sport can be questioned. Thus, athletes expending and eating 5000 kcal/day will ingest twice as much protein as people not involved in exercise and expending/ingesting 2500 kcal/day. Protein intake may thus be sufficient as long as the diet is composed of a variety of foods including lean meats, fish, milk products, eggs and vegetable protein. Nevertheless, supplementation is warranted for athletes who compete in weight classes and combine intensive training with weight reduction programmes, for vegetarian athletes, who consume low energetic and low protein diets (55, 141), or for athletes who for any reason are unable to ingest sufficient protein.

This situation will be found in individuals who ingest a low energetic (weight reducing/maintenance) diet or in those who are not able to meet energy needs by ingestion of normal foods for particular reasons.

This may occur in all categories at risk of marginal nutrient intake as described in chapter 1 (Table 1). Especially when ingesting < 1500 kcal/day, quantitative intake of some essential nutrients may become marginal in relation to the increased needs (23, 176).

Some protein supplementation, to achieve an intake of 1.2–1.8 g/kg body weight a day, the optimal daily requirements for athletes, may be warranted in these situations, to maintain normal nitrogen balance and to reduce impairment in training status. Also, endurance athletes involved in prolonged heavy multiday exercise may benefit from protein supplementation in the form of "soluble" proteins, since these reduce the time of digestion and thus of gastrointestinal bulk.

It should also not be overlooked that meals which replace normal meals during ultraendurance events, such as the triathlon, multiday endurance events, and high altitude climbing, may be composed of CHO, fat and protein in a ratio of 60–70 en% CHO, 10–15 en% protein, 25–30 en% fat. Protein sources used for supplementation or as part of replacement meals, taken during "prolonged endurance exercise days", should be low in fat, easily digestible and of appropriate quality. Milk protein, milk protein hydrolysates and combinations of proteins such as whey protein and/or caseinates seem to be appropriate for these purposes. These protein sources are additionally very low in fat, cholesterol free and do not increase purine intake and uric acid. From a health point of view, these protein sources can therefore replace a substantial part of the high animal protein intake, or textured vegetable protein and tofu-soy protein intake, the alternative protein sources derived from soy beans, sources known to be high in purines. Alternative use of these protein sources will also reduce the atherogenic character of diets in weight training athletes, who consume large amounts of eggs (62).

Although research is currently being conducted on the role of amino acid supplementation during exercise, there is no evidence that supplementation of single amino acids will be of benefit to performance. However, some amino acids have characteristics which may make them interesting for enrichment of sports food products. Branched chain amino acids (BCAAs) are known to pass the liver almost exclusively and may thus be an optimal nitrogen supplier for muscle tissue in periods of recovery when protein synthesis is increased (125, 194). The plasma AA concentration of glutamine is known to be decreased in endurance athletes and has been suggested to be essential for optimal immune competence as well as for protein synthesis (139, 188). Both BCAAs and glutamine may therefore be useful for supplementation in specific conditions, e.g. repeated exhausting endurance exercise. Further research is needed to support these possible applications. Performance improvement claims have been made with respect to supplementation of the amino acids arginine, ornithine, tryptophan and BCAAs. Effects, if any, are of nutritional origin. Therefore, these amino acids will be discussed in more detail in chapter 5.

3
Aspects of Dehydration and Rehydration in Sport

FLUID AND ELECTROLYTES

Fluid is often forgotten in discussions about nutrient requirements. However, humans can live for a prolonged period of time without macro- and micronutrient intake, but not without water.

Water is the basic substance for all metabolic processes in the human body. It enables transport of substances—required for growth and energy production—by the circulation and interchange or exchange of nutrients and metabolic end-products between organs and the external milieu. Water balance is regulated by hormones and the presence of electrolytes, especially sodium and chloride. In order to explain the importance of water and electrolytes involved in fluid homeostasis for the exercising individual, we shall briefly describe how fluid balance is related to health and performance and how this is influenced by exercise.

Fluid reserves

Water is the largest component of the human body, representing 45–70% of total body weight.

An average 75 kg human "contains" about 60%, or 45 liters of, water. Muscle comprises approximately 70–75% water whereas fat tissue contains only about 10–15% (168).

From this it can be concluded that trained athletes who have a lean body and low fat mass have a relatively high water content. Under normal conditions (adequate fluid intake) the body water content is kept remarkably constant. It is not possible to store water in the body as any excess water will be excreted by the kidneys. On the other hand it is possible to dehydrate the body by having an imbalance between fluid intake and fluid losses.

In such a situation water will be lost from two main compartments in which the water content is normally kept constant.

1. The intracellular compartment.
2. The extracellular compartment.

The extracellular compartment can further be separated into interstitium (space between the cells) and vasculum (space within the blood vessels).

A semi-permeable cell membrane separates the intracellular water from the water that surrounds the cells.

The water content of all compartments is mainly determined by osmotic pressure, caused by osmotically active particles. Due to the semi-permeability of membranes, as well as ion pumping, the concentration of electrolytes in intra- and extracellular compartment differs.

Water itself can freely pass cell membranes. Osmosis is defined as the passage of water from a region of lower solute concentration to a region with higher concentration. The ultimate result of this water shift is to equalize the two solute concentrations. In the human, body fluid shifts are made to normalize extracellular fluids at approximately 300 mOsm (isotonicity). Particles which are osmotically active in the body are mainly proteins, electrolytes and glucose.

Apart from solute concentration, blood pressure also exerts an important effect on fluid exchange. Blood pressure together with osmotic effects determine the rate at which water leaves the circulation to enter the tissues, or enters the bloodstream from the tissues.

A change in one compartment, e.g. pressure or solute concentration, can directly or indirectly influence the fluid/solute status of the other compartments. For example, during the first few hours of water deprivation, fluid is lost mainly from the extracellular compartment. Blood fluid and plasma volume will decrease, resulting in a compensating water flow from the tissue (interstitium) to the blood. With continuing water deficits the remaining tissue water will therefore become increasingly concentrated. This will initiate water loss from the cells, finally resulting in cellular dehydration. Changes in fluid regulatory hormones will stimulate the kidney to reabsorb water and sodium in this circumstance (136). Both extracellular (tissue) and cellular dehydration are known to initiate thirst, a stimulus to ingest water for rehydration (74).

Additionally, severe dehydration will initiate impaired metabolism and heat exchange. Intensive physical exercise, especially when executed in the heat, may lead to dramatic changes in fluid content as well as electrolyte concentration in the different compartments (129, 166, 167, 168).

Intracellular fluids and electrolytes

Total intracellular fluid content amounts to approximately 30 liters, about two-thirds of the total body water. This amount of

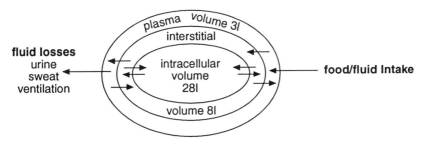

Figure 22 *Representation of different water compartments in the body, as well as their fluid exchange routes*

water is primarily kept within the cells by an osmotic drive caused by the relatively high electrolyte and protein content. An approximate concentration of electrolytes in intracellular fluid is given in Table 2. Sodium and chloride (outside the cells) and magnesium and potassium (inside the cells) are the most important electrolytes exerting an effect on cell water content.

Influence of exercise

Muscular contractions will result in the production and accumulation of metabolic end-products inside the cell. Initially these end-products will cause an osmotic gradient leading to a net uptake of water into the cell.

At the same time transport processes are initiated and changes in membrane permeability take place. These will lead to transfer of metabolites and potassium from the inner to the outer side of the cell. As a result, the interstitial water will become hypertonic (more concentrated) compared to blood with the result that water will shift from the blood to the interstitium. Increased blood pressure will further favor this shift (42, 74, 166, 167).

Table 2 *Approximate concentration (mEq per liter) of electrolytes in the intracellular fluid, interstitial fluid and plasma (166)*

	Intracellular (Skeletal Muscle)	Extracellular Plasma Water
Cations		
Sodium	10	130–155
Potassium	150	3.2–5.5
Calcium	0	2.1–2.9
Magnesium	15	0.7–1.5
Anions		
Chloride	8	96–110
Bicarbonate	10	23–28
Organic Phosphates	65	0.7–1.6

As a result, plasma volume will decrease immediately by ±10% after the onset of exercise and this decrease will slowly return to a lower level of 3–5% thereafter, unless dehydration takes place, causing secondary hemoconcentration (166).

Muscle volume increases during exercise as a result of fluid shifts into skeletal muscle. This increase is most pronounced during high intensity anaerobic work, which causes a large intracellular lactic acid production and accumulation.

The water pool "between" blood and intracellular space may be "stressed" from two sides during exercise. On the one hand, muscle cells will take up water as described above. On the other hand, if sweat losses are large, causing plasma volume to decrease and blood electrolyte levels to increase, then the blood will start to draw water from the interstitial space, too. As a result, interstitial water content will fall to low levels.

Finally, if this situation continues, the whole process described initially will become reversed, leading to intracellular dehydration (74, 166, 168).

Extracellular fluid and electrolytes

As described before, the extracellular space can be divided into two subcompartments:

1. The interstitium, the space surrounding the cells and making up the interstitial fluid.
2. The vasculum, the space within blood vessels, for blood plasma.

Total water content of these compartments is approximately 11.5 and 3.5 liters respectively, giving a total of 15 liters extracellular fluid, equal to 50% of intracellular fluid (166). The interstitial fluid is the exchange medium between the cells and the blood.

The blood is the final transport medium to deliver O_2 and nutrients to the tissues and to transport water and metabolic end-products such as lactate, ammonia and CO_2 to the lungs, liver, kidneys and skin for elimination and/or excretion.

Regulation of fluid and electrolyte homeostasis, by means of the excretion/retention processes in the kidney, is subject to complex hormonal stimuli (186) The approximate electrolyte concentrations of these two subcompartments are also given in Table 2. Major differences in electrolyte concentrations exist with potassium and sodium. Potassium is the major intracellular ion. Sodium and chloride are the major extracellular ions. Therefore, sodium and chloride can be regarded as the most important osmotically active electrolytes.

Influence of exercise

The water content of the muscle tissue will increase and blood plasma will decrease, due to repeated muscle contractions.

With continuous exercise the water content of all compartments will further decrease as a result of fluid loss by sweating and insensible water loss from the lungs, especially at high altitude. Depending on exercise intensity, training status, climatic circumstances and body size, sweat losses may range from a few hundred milliliters to >2 liters per hour (32, 196).

Because plasma fluid is of prime importance to maintain a normal blood flow through "exercising tissues", it may be concluded that a dramatic decrease in plasma volume will lead to decreased blood flow. This will automatically lead to reduced transport of substrates and oxygen to the muscles, needed for energy production, and of metabolic waste products, including heat, from the muscle to eliminating organs.

The first may lead to a decreased energy production capacity and will induce fatigue. The latter will lead to a decreased heat transfer from the muscles to the skin resulting in increasing core temperatures (32, 114, 167, 168).

Figure 23 *Ultraendurance competitions in the heat may be of risk to health. Minimal clothing in bright colors and regular fluid intake are needed to minimize heat stress*

In particular endurance athletes exercising in the heat may be prone to dehydration → heat exhaustion → heatstroke/collapse (129, 167, 168, 181). Metabolic water production during endurance exercise may be significant, but is insufficient to compensate for fluids lost through sweating.

The electrolyte concentration of sweat is lower than that of blood. This means that relatively more water than electrolytes is lost from the blood. (Sweat electrolyte concentrations are given in Table 3.)

Therefore, dehydration due to sweat loss will lead to an increase in the concentration of blood electrolytes (114). However, this is only the case when no water is ingested to compensate for the fluids lost. Large sweat losses and compensation by plain water intake may induce too low blood sodium levels.

Hyponatremia and consequently signs of water intoxication have been observed in marathon runners and triathletes (130,

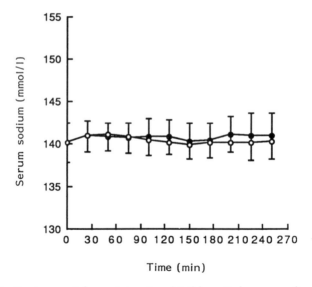

Figure 24 *During a 4 hour intensive biathlon (3 hours cycling, 1 hour running) plasma sodium was measured in eight elite triathletes ingesting plain water (—○—) and eight ingesting an isotonic glucose electrolyte drink containing 600 mg sodium/liter (—•—). Total fluid intake was 600 ml/hour. No effect of sodium intake on serum sodium was observed. Taken from data of Brouns (214)*

Table 3 *Electrolyte content of whole body wash-down sampled sweat derived from the data of 13 studies*

Electrolyte	Cl^-	Na^+	K^+	Ca^{2+}	Mg^{2+}
Average (mmol/l)	28.6	32.7	4.4	1	0.79
SD	13.5	14.7	1.3	0.7	0.6
Average (mg/l)	1014	752	173	40	19
SD	481	339	52	27	15
Range (mg/l)	533–1495	413–1091	121–225	(13–67)	(4–34)
Bioavailability	100%	100%	100%	30%	35%
Correction factor	0	0	0	×3.33	×2.86
Proposed replacement range (mg/l)	500–1500	400–1100	120–225	45–225	10–100

For the principal electrolytes the table represents 274 observations made on 123 subjects. Net absorption in the gut is assumed to be 100% for Cl^-, Na^+ and K^+ and 30% and 35% for Ca^{2+} and Mg^{2+} respectively. Thus, replacement of the electrolytes lost requires equal amounts of Na^+, Cl^-, and K^+ but larger amounts of Ca^{2+} and Mg^{2+} Taking a correction for absorption into account reveals an upper replacement level (32, 212). Reproduced with permission from Chapman & Hall, London

142, 143, 144). The argument that the sodium content of the meals, ingested post exercise, is enough to compensate for the losses, is misleading in this context, since post exercise meals do not compensate for losses during exercise. For this reason and because of its positive relation to glucose and with it water absorption, sodium is recommended in dehydration drinks for athletes (see section 3.1.2).

Regular endurance training resulting in large sweat responses will lead to adaptations in favor of a better maintenance of fluid and electrolyte balance. Sweat glands will adapt to reabsorb sodium and plasma volume tends to increase. Also the sensitivity for fluid regulatory hormones will be enhanced (114, 129, 196). Sweating will become more "economical and effective". Less sweat will drip off the body.

Nevertheless, trained people exercising at their maximal levels of endurance performance capacity will be prone to dehydration during competition or training. The thermogenic stress, caused by the extremely high metabolic rates, will initiate maximal sweat rates.

Fluid and electrolyte intake

Daily fluid intake is normally associated with food consumption (salty/spicy foods) or having a dry mouth. To a large extent this accounts for learned (conditioned) drinking behavior. True thirst, however arises as a consequence of intra- and extracellular dehydration (74).

In general, fluid intake should equal total daily water turnover which is assumed to be ±4% of body weight in adults (131). Total water turnover can vary markedly, mainly because of differences in metabolic rate (exercise will greatly influence this factor) and in insensible water loss. The latter can be strongly influenced by climatological circumstances as well as by altitude. Acute water loss in large quantities can also result from diarrhea.

Because the daily water requirement represents the amount necessary to balance insensible losses (via breathing and skin) and to supply the kidneys with the minimal amount of fluid needed for excretion of metabolic end-products and electrolytes, it is impossible to recommend a general water intake quantity. However, a minimum fluid intake of 1.5–2.0 l/day (for a 70 kg male) may be needed to avoid abnormal metabolic effects.

An intake level of 1 ml/kcal energy expenditure seems a general recommendation (131). Cycling a mountain race in the Tour de France, while expending 6000 kcal/day may then require at least 6 liters of fluid. A level of 6 liters fluid intake has been reported under these circumstances (165). Running a marathon (energy cost ± 3000 kcal, (137)) would then cause an extra fluid requirement of 3 liters. The *minimal* daily requirements for adults given by the National Research Council (1989) for sodium, chloride and potassium, the major electrolytes active in water homeostasis and also lost by sweating, are 500, 750 and 2000 mg respectively.

Daily food intake generally leads to much higher intakes than these figures. Therefore, supplementation is not advisable. However, in the case of substantial losses such as during acute diarrhea or as a result of prolonged intensive sweating,

Figure 25   *Fluid post in the 67 m Swiss Alpine Marathon of Davos. Fluid and carbohydrate are required to maintain optimal performance*

electrolyte levels in plasma may be threatened. In these cases it is advisable to include some electrolytes in rehydration solutions.

Rehydration solutions

Rehydration solutions for athletes are generally designed to replace fluid and minerals lost by sweating and also limited amounts of energy in the form of CHO. All three substances are lost/used during endurance exercise and are subject to mutual influences.

Higher exercise intensities require a higher degree of energy production for which CHO as energy source is most suitable. Accordingly, with higher exercise intensities, more metabolic heat will be produced. Consequently sweat production/loss will be increased, as will the excretion of electrolytes. The longer the exercise the larger the amount of fluid, electrolytes and CHO needed to replace the losses.

There are large differences between individuals in sweat rate, sweat electrolyte content, degree of CHO utilization, etc. (see specific paragraphs). These differences can be further influenced by climatological circumstances. As a result, it is impossible to create a rehydration solution which will exactly compensate for the losses of any individual in any situation. Rehydration solutions for athletes are generally designed to cover the needs of a large exercising population under different circumstances. This is necessarily a compromise which has to be made by any producer. General guidelines for the composition of rehydration solutions have recently been obtained from a large number of studies in the field of gastric emptying, intestinal absorption, fluid balance regulatory factors and fatigue/performance and have been summarized in a number of excellent reviews (30, 32, 40, 42, 69, 113, 114, 126, 130, 153, 155).

The general outcome from these studies is that addition of small to moderate amounts of CHO and sodium to a drink does not delay gastric emptying and improves absorption, compared to plain water. The scientific rationale behind these findings is the fact that coupled glucose—sodium transport across the gut membrane is very fast and stimulates water absorption due to the osmotic action of these solutes (69, 114, 131).

The addition of other electrolytes, in small quantities as lost by the whole body sweat, does not influence gastric emptying, nor absorption (153, 154). The CHO fraction will additionally contribute to the maintenance of a normal blood glucose level and will lead to a sparing of the endogenous CHO reserves (49, 75, 124, 148). The latter may influence protein breakdown, delay fatigue and thus influence performance (25, 43, 44, 45, 113, 127, 188, 189).

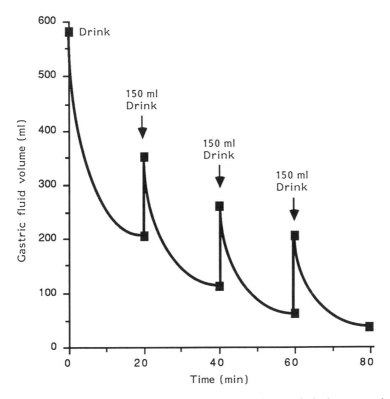

Figure 26 *Gastric emptying rate after ingestion of a single bolus (600 ml) of an isotonic carbohydrate (7%)–electrolyte solution or with repeated drinking as usual in endurance events. Repeated drinking with 70 g CHO/liter does not lead to fluid accumulation in the stomach. Taken from data of Rehrer* et al. *(215)*

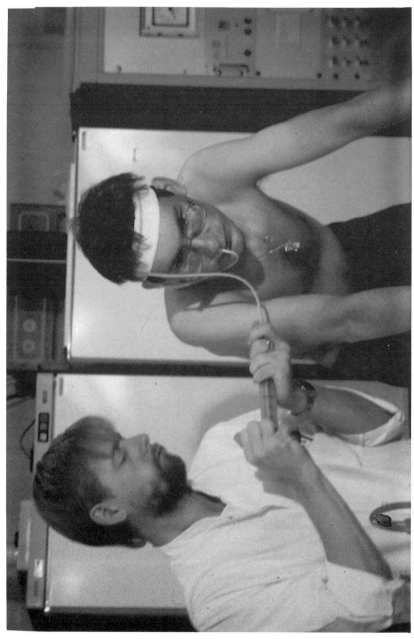

Figure 27 *Determination of gastric volume and gastric secretion by using a nasogastric tube during cycling exercise*

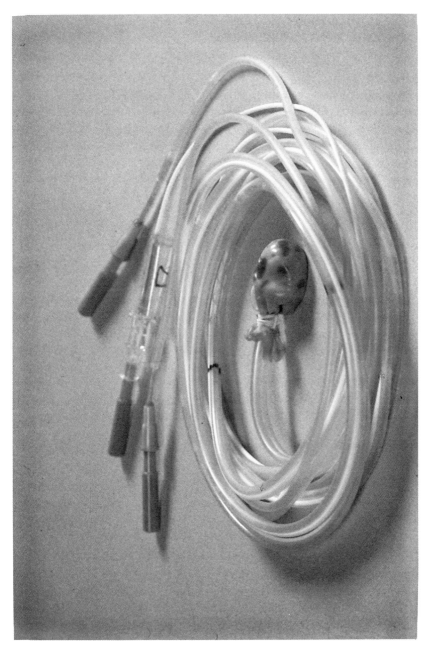

Figure 28 *A triple lumen catheter, which is used to study fluid and substrate fluxes in the jejunum*

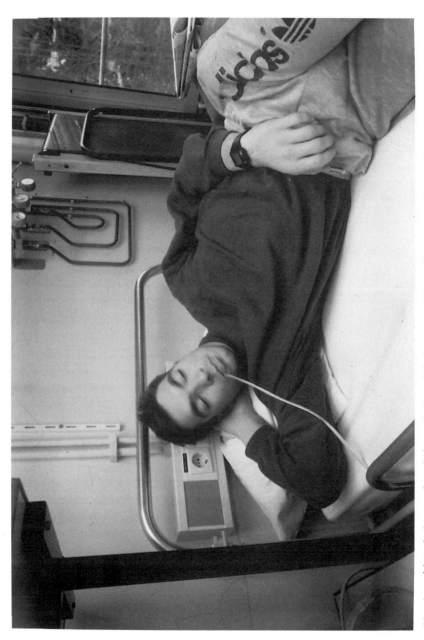

Figure 29 *An athlete being perfused through the triple lumen catheter. These complex perfusion experiments generally last 8–10 hours*

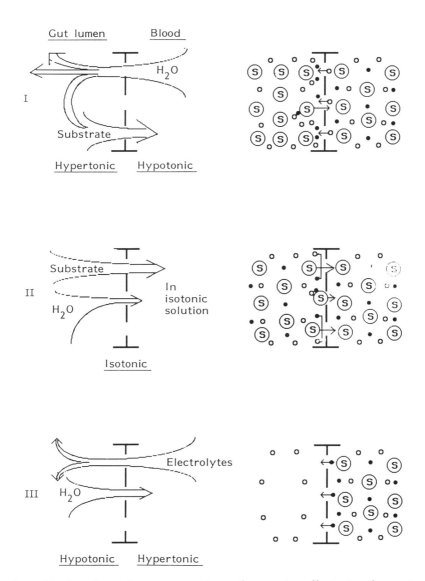

Figure 30 *A schematic representation of osmotic effects in the gut. Water perfusion will induce electrolyte secretion and water-electrolyte absorption. Hypertonic perfusion will induce water secretion and water-substrate absorption. Isotonic perfusion will induce substrate-water absorption. (Net absorption = absorption – secretion.) ○, water; •, electrolyte; S, solute*

A large body of scientific evidence shows that different types of CHO in amounts of 30–80 g/l and sodium in amounts of 400–1100 mg/l induce a high rate of gastric emptying and fluid absorption (69, 114). A maximal fluid absorption rate seems to be a prerequisite only in the event that the quantity of fluid ingested balances or exceeds the quantity that can be absorbed at the same time (for example in the case of massive fluid loss in watery diarrhea). However, fluid intakes during exercise generally do not exceed 600 ml/h in runners or 800 ml/h in cyclists (143), which is much less than the amount which could be absorbed maximally.

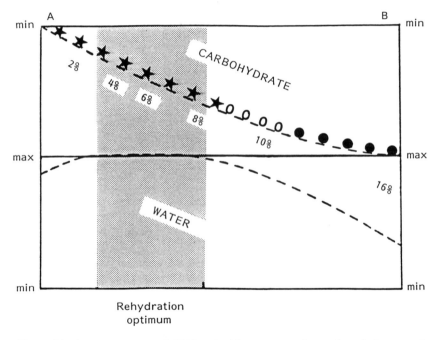

Rehydration
optimum

Figure 31 *Low amounts of CHO stimulate water absorption (left part of figure, "A"). High amounts of CHO in a beverage reduce gastric emptying and induce fluid secretion leading to a reduced net fluid absorption (right part, "B"). "A" leads to high fluid–low CHO availability. "B" induces high CHO–low fluid availability. Maximal CHO availability, without impairing fluid homeostasis is found with beverages containing 60–80 g of CHO/liter. Optimal choice of a drink depends on climatological circumstances and physiological characteristics of the sports event. Reproduced from Brouns (32), with permission from Chapman & Hall, London*

Therefore, it is still open for discussion whether a maximal rate of gastric emptying and absorption is always necessary for the exercising individual. Thus, slightly more concentrated CHO–electrolyte solutions (up to 100 g CHO/l) are known to reduce net fluid absorption only slightly, but enhance CHO availability. In the case of submaximal fluid intakes such drinks do not seem to differ in effects on fluid homeostatis, compared to water or very diluted CHO solutions (32, 40, 114, 126, 127).

Flavored drinks are preferred by athletes to plain water. Consequently such drinks are ingested in larger volumes (88).

A general guideline should be that rehydration solutions should not be strongly hypertonic (i.e. <500 mOsm, preferably ≤330 mOsm). Hypertonic solutions have been shown to reduce net fluid absorption by inducing fluid secretion into the gastrointestinal tract, to achieve isotonicity with blood, and may also reduce gastric emptying. The latter may influence/limit quantitative fluid intake (30, 32, 114, 115, 155).

The source of CHO will influence fluid osmolality. Therefore, so as not to result in very high osmolalities, the quantity of monosaccharides dissolved must be smaller than that of disaccharides or polysaccharides.

Based on the current knowledge and evidence, a general recommendation for the composition of oral rehydration beverages for sport is given in Table 4.

Table 4 *Oral rehydration soluations for combined fluid carbohydrate–electrolyte supply in sports*

Recommended		Optional	
Carbohydrate	30–100 g/l[b]	Chloride[a]	max. 1500 mg/l
Sodium[a]	max. 1100 mg/l	Potassium[a]	max. 225 mg/l
Osmolality	<500 mOsm/l[c]	Magnesium[a]	max. 100 mg/l
	favorable ≤isotonicity	Calcium[a]	max. 225 mg/l

Carbohydrate sources:	Maximal amount of CHO (to avoid hypertonicity and/or a too high concentration[b])
Fructose	35 g
Glucose	55 g
Sucrose	100 g
Maltose	100 g
Maltodextrins	100 g
Soluble starch	100 g

[a]Quantities taken from Table 3.

[b]Water absorption becomes maximized with approximately 30 g CHO/liter. This is also about the minimum amount of CHO needed to achieve measurable effects on glucose/energy metabolism. The upper level (100 g) is given because gastric emptying rates and therefore fluid availability will be reduced at higher concentrations. Additionally the osmotic load of drinks containing more than 100 g will be increasingly effective in reducing the net fluid absorption. More concentrated solutions cannot be considered as rehydration drinks, but are energy (CHO) supplements.

[c]Net water absorption in the gut, after gastric emptying, is mainly determined by substrate absorption—which pulls water along, and by osmotic gradients. An increase in solute (carbohydrate) concentration will lead to a higher solute absorption and with it water absorption. An increase in osmotic load, however, enhances osmotic fluid secretion into the gut. Net fluid absorption results from two opposite water fluxes (absorption–secretion). Thus, hyperosmolality will counterbalance water absorption benefits achieved by solute transport. Osmolalities of >500 mOsm/l should be avoided.

[d]Fructose as sole CHO source may induce gastrointestinal distress at concentrations of >35 g/l. This is not the case in combination with other CHO (e.g. sucrose)

4
Sport Nutritional Aspects of the Micronutrients

MINERALS

Minerals are essential substances for the musculoskeletal system as well as for numerous biological actions. For example, growth requires minerals as building substances and an insufficient supply of calcium and phosphate is associated with impaired skeletal development.

Minerals are important for nervous transmission processes, muscle contraction, enzyme activity, etc. In the previous chapter on fluid and electrolytes, the role of sodium and chloride in fluid homeostasis was described. Here we shall briefly describe how other key minerals are involved in important functions for the exercising individual and how their requirements are influenced by exercise.

The minerals to be discussed are:

- Potassium.
- Magnesium.
- Calcium.
- Phosphorus.
- Iron.
- Zinc.

Mineral reserves

The mineral content differs among tissues as well as between intracellular and extracellular compartments.

For example, bone has a very high calcium and phosphate content, the muscle cell has a high content of potassium and magnesium, and blood and interstitial water are high in sodium and chloride. Although minerals are components of tissues such as bone or muscle, this does not necessarily mean that they are freely available, i.e. are part of a functional, metabolically available mineral reserve.

Since most minerals are bound to or are a part of functional structures/systems, the amount of metabolically available minerals in fact is very small. The major fraction of the "metabolic" mineral pool is present in blood plasma and interstitial fluid.

The amount of minerals in circulating body fluids depends on the input (from food), on the one hand, and on uptake or release by tissues or losses/excretions (by sweat, urine, feces) on the other.

The content of minerals within body fluids remains within a narrow range. Therefore, any excess of minerals will be compensated by increased excretion. Any shortage will, in the first instance, be compensated by reduced excretion or/and by increased release from tissues, but with continuing shortage plasma mineral levels will start to be reduced. The latter will influence the uptake or release of minerals by the cells and thus the cell mineral status. A period with a relative shortage of one or more minerals, like during a marathon run, will not necessarily mean that health and/or performance are threatened.

However, during prolonged periods of mineral deficits, cell growth and cell function will be impaired.

Potassium

Potassium is the major intracellular cation occurring in cell water at a concentration of about 40 times the concentration of extracellular water (see Table 2). Potassium is important for the transmission of nerve impulses, membrane potential and hence muscle cell contraction, and maintenance of normal blood pressure.

Ninety to one hundred percent of ingested potassium is absorbed in the gut and enters the circulation (131). Plasma potassium content has been shown to influence contractility of both heart and skeletal muscle. Excessive plasma potassium levels produce typical ECG changes and may even lead to a sudden heart standstill. Therefore, large intakes of potassium, leading to excessive blood potassium levels, should be discouraged (131). Potassium is excreted from the body in urine and to a small degree in feces (diarrhea is known to result in high potassium losses) and/or sweat.

Influence of exercise

During repeated muscle contractions potassium is lost from the muscle. This loss is caused by changes in cell permeability and the frequent inward and outward fluxes of sodium and potassium which are part of the electrochemical contraction processes (121, 185).

Additionally, potassium is stored together with glycogen in the muscle fibers (18). Breakdown of glycogen will lead to a liberation of potassium and may subsequently enhance potassium loss from the muscle cell.

As a result, the potassium concentration in interstitial fluid, as well as in blood plasma, will increase. This increase will be most pronounced during maximal exercise intensity (121, 185).

In addition potassium may be lost from damaged muscle fibers, but no evidence for this is available. Muscle fiber damage occurs due to mechanical stress, primarily during activities involving negative work, such as downhill walking/running (6).

Sweat losses incurred during exercise, will result in only small potassium losses. The concentration of potassium in sweat is about equal to that in blood plasma. Post exercise potassium is excreted in larger quantities in the urine (most probably because the kidney is stimulated to retain sodium for fluid homeostasis and will therefore exchange sodium for potassium) (30, 114). Some concern has arisen in the past about the possible effect of sustained exercise and sweat loss on plasma potassium concentration as well as potassium balance.

However, since continued exercise leads to a continuous efflux of potassium from muscle, plasma potassium has not been shown to fall. Any deficit will therefore occur in the intracellular potassium levels. The latter are difficult to measure.

However, intracellular potassium losses may be more than compensated by potassium which is released by the breakdown of intracellular glycogen. In this case there would be no change in intracellular free potassium. Increased potassium requirements *during exercise* are, therefore, unlikely.

However, after exercise potassium requirements may be enhanced. Immediately after exercise there is a very rapid uptake of potassium by the cells. Further, muscle glycogen synthesis and coupled potassium storage proceed at a maximal rate immediately after exercise. As a result, plasma potassium levels are known to decline very rapidly after the finish of exercise to normal resting levels or slightly below (121, 185).

Potassium intake

The recommended minimal daily intake for potassium is 2 g/day (131). However, this figure does not take into account possible losses through sweat. Desirable intake, therefore, is 2–3.5 g/day (52, 131). Potassium is widely available in foods, since it is an essential constituent of all living cells, especially fruits (bananas, oranges), vegetables (potatoes) and meat. Potassium intake may vary considerably depending on food selection. High intakes of particular food items may lead to a high potassium intake, as large as 8–11 g per day (131).

Figure 32 *Tropical fruits and tomatoes as well as their juices have a high content of potassium*

Magnesium

The magnesium content of the body is approximately 20–30 g. About 40% of this amount is located within the cells (especially muscle), about 60% in the skeleton and only 1% in extracellular fluid (155).

Magnesium is an essential mineral present in about 300 enzymes necessary for biosynthetic processes and energy metabolism.

Magnesium plays an important role in neuromuscular transmission and activity: it acts at some points synergistically with calcium, while at others it is antagonistic. Magnesium content of blood plasma is kept within a narrow range. Practically all metabolically available magnesium is located in the very small extracellular pool (see Table 2). Any change in this pool is caused by nutritional intake, by uptake in or release from tissues or by losses or excretion (37, 114). Magnesium absorption in the gut is approximately 35%. Magnesium is excreted mainly by the urine. Feces also contain magnesium. However, this represents principal unabsorbed magnesium. Magnesium is further lost through sweating (see Table 3).

Influence of exercise

The extracellular magnesium pool represents the principal fraction of what is metabolically available. Any uptake or loss from this pool will, therefore, result in a lower magnesium concentration.

Low resting as well as exercise plasma magnesium levels have repeatedly been reported in athletes involved in regular endurance exercise. However, several different explanations have been given for this decrease.

It has been suggested that this decrease results from magnesium loss through sweat as well as from an uptake by red blood cells and fat cells. Losses through sweat are generally small (see Table 3) but may become significant with prolonged sweating. Therefore, it is difficult to decide whether a reduced plasma magnesium level in athletes represents a true marginal (deficit) status or whether this is simply a result of physiological magnesium shifts. Additionally, magnesium loss may be increased for 24 hours after strenuous exercise. Marginal magnesium availability has been associated with impaired energy metabolism, greater fatigue and the occurrence of muscle cramps (37, 131), although the latter could not be confirmed in a study on marathon runners (211)

Magnesium intake

The recommended daily intakes for minerals other than sodium, potassium and chloride are given in Table 5. The data given in this table represent the quantities established by expert panels in the USA and Germany. These quantities are thought to be adequate for sedentary people. As yet there are no guidelines for athletes, who may have higher daily requirements for most nutrients.

The magnesium content of food varies widely. Fish, meat and milk are relatively poor in magnesium, while vegetables, exotic fruits, berries, bananas, mushrooms, nuts, legumes and grains are relatively rich.

Table 5 *Recommended dietary allowances for minerals (mg)*

Age	Magnesium	Calcium	Phosphorus	Iron	Zinc
Males					
15–18/15–18	400/400	1200/1200	1200/1600	12/12	15/15
19–24/19–25	350/350	1200/1000	1200/1500	10/10	15/15
25–50/25–51	350/350	800/900	800/1400	10/10	10/15
Females					
15–18/15–18	300/350	1200/1200	1200/1600	15/15	12/12
19–24/19–25	280/300	1000/1000	1200/1500	15/15	12/12
25–50/25–51	280/300	800/900	800/1400	15/15	12/12

The data given in milligram (mg) are derived from NRC/DGE.
NRC = National Research Council, Recommended Dietary Allowances, 1989 (USA).
DGE = Deutsche Gesellschaft für Ernährung, Empfehlungen für die Nährstoffzufuhr, 1991.

Magnesium intake has been found to decline over recent decades, most probably due to the increased consumption of refined and processed foods. For example, >80% of the magnesium found in unmilled grain is lost by removal of the germ and outer layers (131). Data concerning magnesium intake in athletes are scarce. A recent study, however, indicated a close relationship between magnesium intake and energy intake (59). Endurance athletes in particular, having higher daily energy intakes, were observed to have adequate daily magnesium intakes compared to the daily recommended allowance for sedentary people. The latter, however, may not represent optimal quantities for athletes, since magnesium losses with urine and sweat are increased as a result of intensive training.

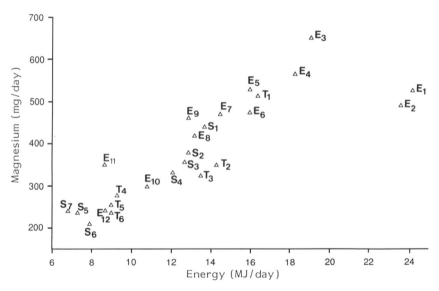

Figure 33 *Magnesium intake in athletes increases with higher energy intakes. E, endurance athletes, S, strength athletes, T, teamsports athletes. From Erp-Baart et al. (59) and Erp-Baart (60)*

Calcium

The human body contains about 1200 g of calcium of which approximately 99% is built into the skeleton. Only a fraction

(1%) is present in extracellular fluid and intracellular structures of the soft tissues (131).

This small fraction represents the metabolically available pool. Plasma calcium is maintained in a narrow range mainly by hormones which control absorption, secretion and bone metabolism. Calcium entering plasma is derived from food or from release from the bones. Calcium is lost through urine, sweat and feces. However, the calcium present in feces mainly represents unabsorbed calcium. In adults net calcium absorption in the gut amounts to approximately 30% (131). Urinary excretion is largely influenced by food intake. Urinary calcium excretion increases with higher protein intake, especially if phosphorus intake remains at the same level (110, 131).

Bone is constantly turning over, thus constantly absorbing and releasing calcium together with phosphate. When calcium intake is too low, plasma calcium levels will remain constant, at the cost of reduced absorption into the bone and normal or increased release from bone. Therefore, the plasma calcium level does not reflect the true calcium status of the body (37).

Influence of exercise

During exercise, calcium plays an essential role in initiating muscle contraction. Calcium liberation within the cell initiates a state of contraction whereas re-uptake initiates relaxation. Plasma calcium has been shown to remain unchanged, decrease or increase during exercise (37, 169). This variation may be attributed to different factors such as water loss leading to concentration, increased release from the bones due to mechanical stress or reduced uptake by the bones due to decreased bone synthesis.

A large body of scientific evidence has recently shown that female athletes may suffer from stress fractures and/or reduced bone density. This "athletic osteoporosis" has been associated with depressed hormones—due to exercise stress—(especially estrogen) which are known to regulate calcium metabolism, as well as relatively low calcium intakes (37). This causes a high frequency of this abnormality among female athletes. Athletes involved in strength training and consuming a high protein diet may excrete more calcium through the urine, especially when phosphorus intake is not increased in parallel to protein intake. Calcium losses through sweat are small (see Table 3).

Calcium intake

Calcium intake also varies widely according to the quantity and composition of the diet. Dairy products form the major source of calcium intake. Nuts, pulses, some green vegetables (broccoli) and seafoods as well as calcium from water may further contribute. Daily calcium intake depends both on food selection and the total food/energy intake.

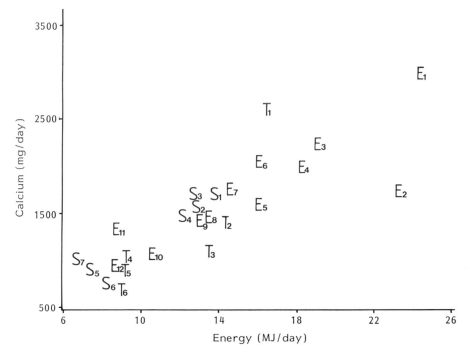

Figure 34 *Calcium intake in athletes increases with higher energy intakes. E, endurance athletes, S, strength athletes; T, teamsports athletes. Reproduced from Erp-Baart et al. (59), with permission from Georg Thieme Verlag, Stuttgart and New York*

Athletes with low daily energy intake or those who follow a weight reduction program may, therefore, be prone to marginal calcium intake.

Females, especially long distance runners, have often been found to have calcium intakes which are lower than the RDA, probably as a result of relatively low energy intakes (37, 59, 84, 131, 206). It has also been reported (87) that a calcium intake of 1500 mg/day is needed to achieve calcium balance in postmenopausal women not receiving estrogen medication. Barr (9) concluded from these data that female athletes who are amenorrheic have similar hormone levels and, therefore, that calcium intake should be 1500 mg/day. On this basis, all amenorrheic athletes (at-risk groups include runners, dancers, gymnasts, bodybuilders) would have inadequate intakes.

Phosphorus

Phosphorus is the counterpart of calcium in bone formation. About 85% of the total body store is present in the skeleton. The remainder is distributed between extracellular and intracellular space in soft tissue.

Phosphorus intake, and consequently supply to the blood, is known to affect bone formation. Therefore, intake of phosphorus and calcium should be balanced. Phosphorus absorption in the gut is approximately 70%, about twice as high as calcium absorption (131).

Phosphorus is excreted in the urine (mainly), feces (mainly the unabsorbed fraction) and in minor amounts with sweat.

Phosphorus is additionally an essential element in numerous enzymes as well as in energy metabolism (nucleotides and conjunction with B vitamins). Virtually all phosphorus (P) circulating in blood and present in tissues exists in the metabolically active form as the phosphate (PO_4^{3-}) molecule.

Influence of exercise

Since exercise results in sweat loss and hemoconcentration, it may enhance plasma phosphate levels. Phosphate losses through sweat are negligible. In addition, changes in alkalosis (inducing a fall in phosphate levels), acidosis and cell damage (inducing an increase in phosphate levels) are known to influence plasma levels (104).

Phosphorus intake

Phosphorus is especially present in protein-rich foods such as milk, meat, poultry and fish as well as in cereal products.

The amount of phosphorus in normal diets is about 1500 mg and has been relatively constant over the years. Increases in daily energy intake will normally also increase phosphorus intake. Therefore, phosphorus deficiencies normally do not occur among healthy exercising individuals (131).

Iron

Iron is an important constituent of hemoglobin, myoglobin and a number of enzymes. As such the, availability of iron is important for oxygen binding and transport as well as for the transfer of electrons in the electron transport chain and for energy production.

About 30% of total iron is found in storage forms as ferritin and hemosiderin and a small part as transferrin. Therefore, these iron stores can serve as indicators of iron status. Poor iron status is indicated by low levels of serum ferritin, increased red cell protoporphyrin levels, reduced transferrin saturation levels and reduced hemoglobin levels. With inadequate iron intake the storage form will be the first to be affected. With prolonged iron shortage hemoglobin production will finally be affected, resulting in iron deficiency anaemia. The latter will reduce oxygen transport capacity and may thus affect endurance performance capacity (131, 145, 150).

Influence of exercise

There is considerable controversy about the extent to which athletes are any more iron deficient than the normal population, especially with respect to hemoglobin concentrations, which are known to be low in many athletes. However, since plasma volume increases as a result of endurance training, the absolute amount of circulating hemoglobin is not necessarily lower but rather an effect of the increased plasma volume. In this case there is a pseudo anemia.

However, over the last 10 years a large body of evidence has indicated that a substantial number of athletes involved in regular training do have decreased iron stores, indicated by reduced bone marrow iron, enhanced iron binding capacity and low serum/plasma ferritin levels.

With respect to the latter, data may be misleading since stressful exercise has been shown to result in temporarily increased levels. Thus serum ferritin levels obtained shortly after intense endurance exercise may no longer accurately reflect body iron stores (102).

Although there is evidence indicating that a part of the poor iron status observed in athletes can be explained by selecting a diet which is poor in (biologically bound) heme-iron— especially in athletes consuming vegetarian and high fiber meals—there is also some evidence that exercise itself can increase heme-iron requirements/losses.

Significant amounts of iron are lost with sweat. This may explain effects of exercise on iron status. Mechanical effects present during the landing phase of the foot while running may lead to red blood cell damage inducing hemolysis and reduced hemoglobin levels. The iron of damaged hemoglobin will enter the circulating iron pool and will then be newly available (37, 56, 57, 118, 132, 145, 150, 152).

The effect of exercise on iron absorption in the gut, especially ultraendurance activities which reduce intestinal blood flow and may affect other transport processes, remains largely

Figure 35   *The footsole is prone to high pressure peaks during the landing phase while running. It is hypothesized that red blood cells with a low stress tolerance may be damaged, leading to anemia*

unknown. Also unknown is the extent to which decreased iron stores in athletes increase iron absorption, as has been observed in sedentary individuals. Recent studies have shown that endurance exercise may lead to gastrointestinal bleeding. As a result, increased fecal hemoglobin and iron loss have been observed (30, 123).

Iron intake

Red meat, liver, poultry, dark green vegetables and cereals (especially iron fortified products) are the major sources of iron intake. Heme-iron in meat is the best absorbable iron source. Iron absorption from non-heme-sources is less and can be further decreased by other food/meal constituents. Vitamin C enhances inorganic iron absorption, while dietary fiber, tea, coffee and calcium phosphate reduce absorption (131). Iron intake in vegetarian athletes is often too low, because of the omission of heme-iron sources from the diet. Female athletes or those who compete in weight class sports or gymnastics often have marginal or inadequate intakes as a result of both qualitative food selection and reduced quantitative energy consumption (38, 55, 84, 176, 206). The recommended/safe daily intake for iron is shown in Table 5. It is assumed that the required daily intake for athletes exceeds this RDA. Research is needed to establish athletes' requirements.

Zinc

Zinc is present in relatively large amounts in bone and muscle. However, these stores are not metabolically available. The body pool of readily available zinc, mainly in blood, is small and has a rapid turnover rate. Zinc is involved in growth and development of tissues, especially muscle, as it is an essential substance in numerous enzymes involved in major metabolic pathways. Recent studies have indicated that zinc may also play a crucial role in immune competence (5, 97, 131).

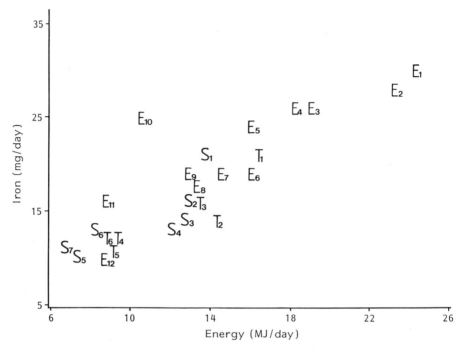

Figure 36 *Iron intake in athletes increases with higher energy intakes. E, endurance athletes; S, strength athletes; T, teamsports athletes. Reproduced from Erp-Baart et al. (59), with permission from Georg Thieme Verlag, Stuttgart and New York*

Influence of exercise

Since plasma zinc represents to a large extent the readily available zinc pool, any rapid change in plasma volume, due to the performance of physical exercise, will affect plasma zinc status either by plasma volume decrease due to dehydration (this will increase zinc concentration due to hemoconcentration) or plasma volume increase in the post exercise phase due to water and sodium retention (this will decrease zinc concentration). Apart from these effects it may be assumed that functional zinc shifts between tissues may occur due to exercise.

It is, therefore, difficult to determine the effect of exercise on zinc status as indicated by plasma zinc levels (5, 37). Plasma zinc increased significantly over several weeks in cycling participants in the Tour de France (164).

Nevertheless, exercise may increase zinc requirements because zinc from the body is primarily lost by urine and sweat and both are enhanced as a result of endurance exercise (4, 5, 37, 41, 78). The increased urine zinc excretion may be related to muscle fiber damage induced by mechanical forces during exercise. Especially in the post exercise phase, zinc may be lost from damaged cells but this is not supported by any solid evidence. Similarly, fasting, leading to muscle loss, has been reported to increase serum and consequently urine zinc levels (178). A very recent study, however, did not confirm effects of muscle damage on serum zinc levels (216).

Zinc intake

Meat, liver and seafood are major zinc sources in the diet. Additional zinc may be derived from milk, cereals and muesli. High CHO foods, especially refined sources, are poor zinc sources. Phytate and dietary fiber are known to reduce zinc absorption (131) and may thus reduce the bioavailability of zinc from cereals and muesli. Daily dietary zinc intake appears to be marginal in many sedentary individuals. Low zinc intakes may be exacerbated in athletes by exercise-induced losses (84, 177). However, zinc intake in athletes is observed to be closely related to total energy intake. Most athletes studied in the Netherlands have higher intakes than RDA for normal sedentary people (60). Vegetarian athletes may be prone to marginal zinc intake, especially when consuming low energetic diets (72, 141). However, there is no good evidence that vegetarian persons, in general, are zinc deficient (55). The recommended/safe daily intake for zinc is given in Table 5.

Mineral replacement and supplementation

From the previous paragraphs it could be concluded that mineral intake in athletes, compared to RDA for normal sedentary people, is sufficient in most cases and that supplementation for healthy individuals, consuming a well balanced diet, containing sufficient amounts of meat, fruits, vegetables, cereal and whole grain products, will not be beneficial.

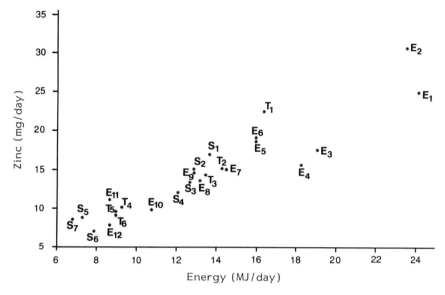

Figure 37 *Zinc intake in athletes increases with higher energy Intakes. E, endurance athletes, S, strength athletes, T, teamsports athletes. Reproduced from Erp-Baart et al. (59) and Erp-Baart (60), by permission of the author*

But, for a number of reasons, the diet of athletes involved in intensive training is often unbalanced. Many athletes consume up to 40% of daily energy intake as in-between snacks, rich in energy, but poor in micronutrients (23, 58, 165). Mineral intake largely depends on both the food selection and quantitative food intake. Thus, there are situations in which supplementation may be desirable, such as any situation where athletes abstain from a normal diet, and during periods of limited food intake combined with intensive training (especially in females, weight class sports participants and in vegetarian athletes; (see Table 1). The addition of minerals to products/meals designed to replace normal meals during ultraendurance events such as triathlon, multiday endurance events and high altitude climbing is recommended; however, the levels should not exceed those of safe daily intake. In this respect it is still an open question whether the RDA established for sedentary people is also adequate for athletes who, due to exercise, may lose substantial amounts of minerals with urine and sweat. Mineral replacement by adding minerals to rehydration drinks is acceptable as long as the mineral levels do not exceed the upper levels reported for whole body sweat (see Table 3). Although substantial amounts of iron can be lost with sweat there is, to our knowledge, no rationale for replacing this mineral in rehydration drinks during exercise.

In general, mineral replacement and/or supplementation in healthy subjects consuming well balanced diets will not enhance performance, but will in certain circumstances contribute to adequate daily intakes. However, in cases of poor food item selection and diet composition, or in vegetarian athletes, athletes may develop impaired status for some minerals especially iron, zinc and magnesium. As well as food education, supplementation may then be advised. In the light of the general goal of athletes to ensure adequate intake in any way possible, the risks of oversupplementation and the invasive/complex nature of determination of mineral status, athletes and their counselors should be educated about dietary practices and safe daily intakes. Micronutrient preparations for athletes should be regulated accordingly; they should not exceed the established levels of safe daily intake.

Some minerals have been promoted for improvement of performance, because of their specific metabolic influences, such as phosphates and sodium bicarbonate. As such, these supplements have nothing to do with mineral supplementation by means of increasing daily intake, compensation of sweat/ urine induced losses or improvement of the mineral status of the body. Because of their use in this way, these substances will be reviewed in more detail in specific sections (see sections 5.7 and 5.8).

TRACE ELEMENTS

The importance of trace minerals (elements) in numerous biological functions, as well as their effect on health and performance, has hardly received any attention until the last decade. This was mainly due to the lack of analytical methods for measuring and evaluating the role of these elements, which are present in body fluids and tissues in "micro-quantities".

Recent technological developments, however, have allowed new insights into trace element status.

Here we shall describe the function and availability of some trace elements and the possible effect of exercise on trace element requirements.

The trace elements to be discussed are:

- Copper.
- Chromium.
- Selenium.

Trace element status

It is only in the last decade that improvements in analytical procedures and techniques have made it possible to study trace elements and their functions in vivo. In exercise science most of the studies have dealt with the macronutrients fat, protein, CHO and water, but utilization, function and storage of these macronutrients is in fact to a large extent regulated by micronutrients (5). Shortage of trace elements in the diet may result in impaired trace element status, which is known to influence biochemical and physiological functions and thereby health. Trace element status is difficult to study. It is possible to obtain samples from serum, tissue, hair, toenails, feces, urine and sweat. Analysis of the first four samples may indicate the status of the pool from which the sample stems. However, this does not necessarily mean that this sample is representative of the whole body or specific tissues. The last three samples may indicate the effect of physical stress on trace element losses and may thus tell us something about the extent of these losses and their meaning for daily requirements. However, increased losses do not give us information on the status of different tissues. The growing knowledge over the last 10 years has indicated different sample sites as representative for different trace elements.

Copper

Copper is an essential element for the human body. Copper deficiency has been shown to result in impaired health and malfunctioning.

Copper is involved in a large number of enzymes, plays a role in energy metabolism, tissue synthesis and protection against free radicals. Additionally, copper influences iron metabolism (131).

Activity of erythrocyte superoxide dismutase (SOD), an enzyme eliminating the damaging effect of free radicals, seems to be an objective parameter for copper status (5, 37, 103).

Ceruloplasmin, the principle copper binding protein present in plasma, may under normal resting conditions give some information on copper status. However, ceruloplasmin is known to be liberated by the liver in larger amounts during periods of stress. The latter may thus lead to wrong information about the true copper status in periods of illness or intensive physical exercise leading to exhaustion (5, 37).

Influence of exercise

Plasma ceruloplasmin as well as serum copper levels have been reported to increase as result of exercise in some studies, but to remain unchanged or to decrease in others. Several factors such as differences in training status, type of exercise, degree of plasma volume change or true copper status may account for this. Copper is lost in significant quantities with sweat (93, 103).

Therefore, it has been suggested that repeated large sweat losses may impair copper status and that an increased dietary copper intake may be required to offset losses by sustained sweating (4, 5). Thus the normal RDA for copper may be too low for athletes. More research is needed on the copper requirements of athletes.

Copper intake

Organ meats, especially liver, are the richest copper sources followed by seafood, nuts, seeds and potatoes. Milk contains only a low level of copper. Observed copper intake in humans is low, 0.9–1.2 mg/day. Zinc, ascorbic acid, iron, calcium, protein, fructose and dietary fiber and also high fructose intakes are known to reduce copper absorption and thus copper status (4). The recommended/safe daily intake for copper is given in Table 6 (131).

Chromium

Chromium acts principally in conjunction (as cofactor) with insulin, potentiating insulin activity. Therefore, chromium is an essential substance for the normal regulation of the blood glucose level. Accordingly, experimental chromium deficiency results in decreased insulin action, impaired blood glucose regulation, or even diabetes.

Because of its role in insulin–CHO–energy metabolism, chromium is thought to be of particular importance for people involved in heavy physical work and consuming CHO rich diets.

Table 6 *Recommended/safe daily intakes for trace minerals*

Source	Copper (mg) f + m	Chromium (μg) f + m	Selenium (μg) f + m
NRC	1.5–3.0	50–200	55–70
DGE	2.0–4.0	50–200	20–100

The data given are derived from NRC and DGE.
NRC = National Research Council, Recommended Dietary Allowances, 1989 (USA).
DGE = Deutsche Gesellschaft für Ernährung, Empfehlungen für die Nährstoffzufuhr, 1991.
f = female, m = male.

Blood chromium does not appear to be a good marker of chromium status. Urinary chromium losses are a cumulative total of small transitory changes in the blood and appear to be a better indicator of changes in chromium metabolism (3, 5, 97, 131).

Influence of exercise

Different stresses, including exercise, infection and physical trauma, are known to exacerbate the signs of marginal chromium deficiency. In the case of exercise this occurs most probably because exercise enhances chromium losses with urine. Additionally, CHO rich diets, especially high glycemia CHO sources such as mono- and disaccharides, are known to increase chromium losses with urine. This is most likely an effect of these CHOs on quantitative insulin secretion and subsequent degradation.

The loss of chromium in sweat has not been quantified using acceptable collection and analytical techniques.

Animal studies have indicated that a poor chromium status is associated with reduced glycogen stores in liver and muscle and that chromium supplementation enhances glycogen storage in this situation. Since endurance performance as well as protein breakdown are associated with CHO availability, it is suggested that sufficient dietary chromium optimizes endurance performance capacity.

In one study chromium supplementation in the form of chromium picolinate increased lean body mass and decreased fat mass, as indicated by anthropometric measures only, in strength athletes. However, care should be taken with the interpretation of these results, as the chromium status in the test subjects was not controlled and the anthropometric measures used are not a very precise way of determining true muscle mass. It was suggested that the potentiating action of chromium on insulin is responsible for enhanced incorporation of amino acids in muscle tissue, increasing lean body mass (and increased resting metabolic rate), and decreasing fat mass (3, 4,

5, 35, 37, 96, 103). More studies are needed to confirm the results obtained by Evans (61).

Chromium intake

The recommended or suggested safe intake of chromium for adults amounts to 50–200 μg/day (131). However, chromium intake in developed Western countries (USA, England, Finland) has generally been found to be lower (5).

Chromium absorption varies from 0.3 to 1.0% for inorganic chromium and from 5 to 15% for organically bound chromium such as in yeast and is inversely related to dietary intake at normal chromium intakes. Important chromium sources are broccoli, oysters, mushrooms, yeast and bran cereals. Additionally, the chromium content of processed food is known to be increased as a result of processing in metal containers/holding tanks. Chromium excretion is known to be increased with high CHO intakes and absorption may be decreased by high fiber diets. Chromium absorption is known to interact (inhibition of absorption) with iron and zinc (5, 131).

Selenium

Selenium forms an essential part of the enzyme glutathione peroxidase which regulates the breakdown of hydroperoxides in conjunction with vitamin E. As such selenium has anti-oxidant properties and plays an essential role in scavenging free radicals which are known to appear increasingly in situations of trauma, stress and also exhausting exercise. Selenium deficits are also thought to affect muscle tissue, resulting in cardiomyopathy and muscular discomfort or weakness (37, 97, 103).

Influence of exercise

Because of its antioxidative function, selenium may help in

preventing lipid peroxidation induced by exercise and thus offset the degree of cell damage, most probably in the active tissues (such as muscle) or tissues which may be prone to decreased blood flow causing local ischemia (gastrointestinal tract). Data are available showing that vitamin E supplementation reduces lipid peroxidation, but no studies have been performed in which the effect of selenium supplementation has been specifically studied.

No data are available on exercise-induced sweat selenium losses (37, 103).

Selenium intake

The recommended/safe daily intake for selenium is given in Table 6. Seafood, kidney and liver are rich in selenium. Grains and seeds have a high selenium content as well which, however, may vary depending on the selenium content of the ground where growth took place (131).

Selenium intake in healthy humans normally seems to be adequate. No data are available on selenium intake in athletic populations, which indicates the need for research.

Trace element replacement/supplementation

Although a well balanced diet containing a variety of fruits, vegetables, grain products, meats and seafood should assure an adequate trace element intake, it may be concluded from the available literature that healthy people—including athletes— often have a low intake of iron, zinc, copper and chromium. These low intakes may cause a poor trace mineral status which may be exacerbated by exercise-induced losses with sweat and urine as well as by low intakes and enhanced losses induced by the high CHO consumption of athletes, especially in endurance events. There are no documented reports demonstrating that the overall trace element status of athletes is significantly different from that of normal sedentary people.

However, several studies suggest that the trace mineral status of athletes may be compromised.

Therefore, athletes should be educated to optimize the selection of foods for their normal meals.

Daily supplementation with a low dose trace element preparation, supplying not more than the recommended daily/safe intake (Table 6, p. 92), can be advised in periods of intensive training or in any situation where athletes abstain from a normal diet such as during periods of limited food intake combined with intensive training (especially in females, in vegetarian athletes and in weight class sports participants; see Table 1, p. 8). The addition of trace elements to products/meals, designed to replace normal meals during ultraendurance events such as triathlon, day endurance events and high altitude climbing, is acceptable, but should not exceed levels of safe daily intake. Although substantial amounts of copper may be excreted with sweat, there is, to our knowledge, no reason to replace these elements with rehydration drinks *during* exercise. In general, trace element replacement and/or supplementation will not enhance performance but may contribute to adequate daily intakes in athletic populations (5, 37, 84). In view of the general goal of athletes to assure an adequate intake in any way possible, the possible risks of over-supplementation and the invasive/complex nature of determination of mineral status, athletes and their counselors should be educated about dietary practices and safe daily intakes. Micronutrient preparations for athletes should be regulated accordingly; they should not exceed the established levels of safe daily intake.

VITAMINS

Vitamins are essential nutrients for the human body. Vitamins are involved in almost every biological function. They serve as coenzymes in many energy producing reactions, are involved in protein metabolism and cell synthesis, and act as antioxidants.

Here we shall briefly describe the most essential functions of the individual vitamins as well as their role in exercise metabolism and their influence on exercise capacity.

Vitamin status

Several methods are used to determine the vitamin status of the body. Vitamins function in specific metabolic processes. Therefore, any real deficit will affect these processes and may lead to abnormalities or illness.

It is possible to register the occurrence of illness symptoms. However, such symptoms are generally seen as the last stage of vitamin status impairment. Development of analytical techniques has made it possible to study biochemical deficits, i.e. a marginal deficit which will influence the metabolic function of a vitamin before clinical signs become apparent. These measurements include the determination of plasma vitamin levels by high pressure liquid chromatography (HPLC) and enzymatic stimulation tests. Any determination of vitamin status by these methods is invasive and expensive. It is important, therefore, that athletes achieve a level of vitamin intake which will assure an optimal vitamin status. This will eliminate the need for invasive tests. Factors which influence vitamin status are food intake and vitamin density of the food, bioavailability (which means the ability to be absorbed) and losses from the body. The influence of exercise on these factors will be discussed briefly.

Individual vitamins and influence of exercise

In the following paragraphs we will briefly discuss the individual vitamins and reported effects of exercise.

Vitamin B1 (thiamine)

Vit B1 plays an important role in the oxidative decarboxylation of pyruvate to acetyl CoA, an essential step in energy production from CHO. For this reason the requirement for this vitamin has been related to total energy expenditure and to CHO intake. The RDA is set at 0.5 mg/1000 kcal (131). It is accepted now that the vit B1 requirement of athletes may be slightly higher due to increased energy and CHO metabolism. Impairment of maximal oxygen uptake resulting in increased CHO metabolism/lactate production was shown in humans receiving a deficient diet (10). Low intakes of this vitamin as well as biochemical deficiencies have been reported in athletic and sedentary populations, especially in cyclists who consume large amounts of CHO solutions with refined CHO sources (13, 28, 58, 84, 165). There are no controlled studies available on the effect of vit B1 supplementation on performance.

Vitamin B2 (riboflavin)

Vit B2 is involved in mitochondrial energy metabolism. The National Research Council relates B2 intake to energy intake. The recommended daily intake is 0.6 mg/1000 kcal, although it is stated that there is no evidence that the requirements increase with increased energy metabolism (131).

Few studies have shown that B2 requirements for people involved in physical exercise may be increased (28). However, there are no studies indicating low intakes of this vitamin in athletic populations. Studies in which vit B2 was supplemented in top swimmers did not show any effects on performance (13, 58).

Vitamin B6 (pyridoxine)

Vit B6 plays an important role in protein synthesis. For this reason this vitamin is often assumed to be of crucial importance to strength athletes. However, there are no solid data available indicating an increased requirement for athletes. Accordingly, studies in which B6 was supplemented did not improve performance. Some studies indicated performance improvements after supplementation with combined preparations, also including citric acid cycle intermediates. However, it is likely that any effect observed was not due to B6 but to accompanying substances (13). A dietary B6 ratio of 0.016 mg/g protein intake appears to ensure acceptable values for B6 status in adults of both sexes. The RDA is set at 2.0 mg/day for males and 1.6 mg/day for females (131). Recent data are available, indicating insufficient intakes of B6 in different athletic populations (58).

Vitamin B12 (cyanocobalamin)

Vit B12 functions as a coenzyme in nucleic acid metabolism and influences protein synthesis. Vit B12 is surprisingly often used by endurance cyclists and strength athletes because it is believed that this compound can have an analgesic effect on muscle soreness when used in mega doses. Williams (199) and van der Beek (11) reviewed the literature up to 1985 and concluded that there is no evidence for any benefit of supplementation, a conclusion shared by others in more recent reviews. Both oral and parenteral supplementation did not influence any performance related parameter (86, 131). The RDA is 2.0 μg/day (131). No data are available on B12 intake or deficits in athletic populations. Deficits of this vitamin may occur if there is impaired absorption due to lack of gastric factor, or with no meat consumption (only source for B12), as for example in vegetarians.

Niacin

Niacin functions as a coenzyme in NAD (nicotine adenine dinucleotide), which plays a role in glycolysis and is needed for tissue respiration and fat synthesis. The amino acid tryptophan can be converted to niacin. Sixty mg of tryptophan have the same response as 1 mg of niacin and are therefore declared as 1 NE (niacin equivalent). Several authors have hypothesized that this vitamin could influence aerobic power, which is an important factor for endurance performance in athletes (199). However, it has been reported that megadose intake can have adverse effects on performance. This may be induced by the depressing effect of nicotinic acid on free fatty acid (FFA) mobilization. Under exercise circumstances this will enhance CHO utilization, which in turn will lead to a faster rate of glycogen depletion. This has been shown to enhance subjective fatigue and to impair performance (13, 85). The RDA has been set at 6.6 NEs per 1000 kcal or at least 13 NEs at caloric intakes of <2000 kcal (131). No data are available on niacin intake or on deficiencies in athletic populations.

Pantothenic acid (PA)

PA is a component of acetyl CoA, the intermediate citric acid cycle metabolite of CHO and fat metabolism. Williams stated in 1985 that some reports suggested a beneficial effect of supplementation but that conclusive data were not available (199). This has not changed until now. Supplementation with pharmacological doses as high as 1 g/day did not result in any performance improvement (13). The National Research Council concludes that there is insufficient evidence to set an RDA for pantothenic acid. The safe daily intake level is assumed to be 4–7 mg (131). No data are available on PA intake or on deficiencies in athletes.

Folate

Folate functions as a coenzyme in amino acid metabolism and nucleic acid synthesis. The RDA for folate amounts to approximately 3 µg/kg body weight, resulting in a daily RDA of 200 µg for males and 180 µg for females (131). There are no controlled studies available on the effect of folate supplementation on physical performance, nor on folate intake in athletes (13). Plasma folate levels, which may reflect folate intakes, were observed to increase in Tour de France participants, who ingested substantial amounts of vitamin preparations (164). Williams (202) cited recent research that folate supplementation would restore normal folate status to runners who were folate deficient, but did not improve performance capacity.

Biotin

Biotin is an essential part of enzymes that transport carboxyl units and fix carbon dioxide in tissues. The conversion of biotin to active coenzyme depends on the availability of magnesium and ATP. Biotin plays an essential role in CHO, fat, propionate and branched chain amino acid metabolism. Biotin is produced in the lower gastrointestinal tract by microorganisms and fungi. However, the extent of absorption in this lower part of the gut is unknown. There are insufficient data to establish an RDA for biotin. A range of 30–100 μg/day is provisionally recommended as a safe daily intake for adults (131). There are no studies available on supplementation effects, nor on biotin intake in athletes (13).

Vitamin C

Vitamin C (vit C) is probably the most intensively studied vitamin. Vit C is a water soluble antioxidant. It scavenges free radicals which cause cell damage and protects vit E, another antioxidant, from destruction. It participates in many enzymatic reactions by acting as an electron transmitter, and is involved in the synthesis of collagen and carnitine, (the latter is needed for the transport of long chain fatty acids across the mitochondrial membrane). Vit C enhances iron absorption in the gut. It is also needed for the biosynthesis of some hormones (14, 68, 131). Early studies performed during the Second World War showed that insufficient vit C lowered physical performance, and increased the sensation of exhaustion and muscle pain. However, many of the studies performed at that time have now been criticized for their poor methodology, control, and statistical design. More recent, well controlled double blind studies have shown that a state of marginal vit C deficiency does not reduce physical performance in single intensive bouts of exercise There are some indications that vit C may enhance the rate of heat acclimation (13). This may be of benefit to athletes involved in endurance competitions in the heat around different parts of the world. Vit C supplementation did not improve performance in well controlled studies.

In general vit C intake in athletes is sufficient, with the exception of individuals consuming a low caloric diet (12, 13, 28, 58, 68).

Vitamin E (alpha-tocopherol)

Vit E is an antioxidant and scavenges free radicals to protect cell membranes from lipid peroxidation. It functions in concert with vit C, beta-carotene and selenium, and also protects red blood cells from hemolysis (14, 131, 171). In the period 1970–1980, special attention was given to this vitamin after reported beneficial effects of supplementation on oxygen utilization and performance. As with vit C, many of these studies were not well controlled or suffered from poor statistical design. Critical analysis of the literature and more recent results from well designed double blind studies did not bring any solid evidence for performance improvement (13, 14, 84, 171, 206).

More recently, special attention has been given to the anti-oxidant properties since it became possible to measure the effects on free radical pathology. Vit E is able to reduce lipid peroxidation both in animals and humans as measured by penthane exhalation analysis. Free radicals are increasingly produced during high intensity exercise, especially in a situation of relative oxygen deficit. Therefore, it is assumed that vit E may have a protective effect on the exercising individual. Studies at high altitude indicate that vit E can influence metabolic performance parameters and reduce penthane exhalation. However, it is not known which tissues undergo lipid peroxidation during exercise. It may be that the most important sites are the tissues which are prone to ischemia during exercise, such as the gastrointestinal system and the kidneys, but not muscle. A substantial part of muscle cell damage during exercise occurs through mechanical stress on the muscle fibers, especially during negative (eccentric) work such as downhill walking or running. This mechanical stress leads to microruptures which cannot be protected by antioxidants. Accordingly vit E has not been shown to reduce

Figure 38 Vitamin E supplementation may improve oxygen uptake and performance at high altitude (photograph ARPE)

such damage. More research is needed to validate the hypothetical effects of vit E on tissue damage, regeneration and long-term training/performance status particularly at altitude (6, 13, 14, 84).

Vitamins A, D and K

Although the importance of these fat soluble vitamins for health is beyond doubt (131), there are no studies available which indicate any significant effect of these vitamins on biochemical or physiological parameters concerned with physical performance. Vit A and beta-carotene have antioxidant properties. However, limited data are available concerning the effects of these substances in the exercising individual. Since these vitamins are potentially toxic, when taken in high doses for a prolonged period of time (with exception of vit K), and daily intake in Western civilized countries is sufficient, there is no need for supplementation (13, 83, 84, 199). Further research is needed to quantify the effect of vit A and beta-carotene on tissue lipid peroxidation.

Vitamin intake

Some aspects of intake of the individual vitamins in athletes have been discussed in the previous paragraphs. Here we will discuss some general influences on daily vitamin intake. Vitamins are present in a wide variety of fresh unprocessed foods such as vegetables, fruits, grains and other starchy foods. A normal well balanced diet composed of a variety of foods is, therefore, believed to supply all necessary vitamins in sufficient quantities. However, in some situations intake may either be insufficient with respect to RDA, or, when sufficient, may not lead to adequate bioavailability due to impaired absorption. The latter may be caused by gastrointestinal disorders or by the presence of substances which inhibit absorption.

Insufficient supply of vitamins occurs in general when people ingest low amounts of food or an unbalanced diet. This first

Figure 39 *Mountain running leads to mechanical damage of muscle fibers. It is hypothesized that recovery from this damage is improved when antioxidant vitamins are supplemented*

situation occurs frequently in athletes who compete in weight categories and follow weight reducing training programs or in athletes who have to maintain a low body weight for prolonged periods of time such as female dancers and gymnasts (see Table 1). An almost linear relationship with energy intake has been observed for thiamine (58).

Vitamin restoration and supplementation

As discussed in the sections on minerals and trace elements (sections 4.1 and 4.2), individuals at potential risk of marginal micronutrient supply are those who consume <2000 kcal for prolonged periods of time. An insufficient supply of vitamins may also occur when large amounts of processed foods constitute the major part of the daily diet. This has been observed to be the case in endurance athletes who ingest relatively large amounts of refined CHO in solution during their sports events (23, 58, 165). The reason for this has been discussed in Chapter 2 in the section on CHO. In both situations the required micronutrient density (i.e. the amount of vitamins present in 1000 kcal delivering food) is higher than can be achieved with the diet. In these situations athletes can be advised to take vitamin supplements (not more than 1–2 times RDA/daily) to enhance micronutrient density.

In processed products/meals vitamins are often added to replace processing-induced losses (restoration) or to increase the vitamin content above normal (enrichment/fortification). Vitamin enrichment of food products/meals designed to replace normal losses during multiday endurance events and high altitude climbing is acceptable; however, intake should not exceed levels of normal nutrient density and safe daily recommendations.

In general, vitamin restoration of energy dense processed foods, or supplementation with preparations, will not enhance performance (13, 195) but may, in athletic populations, contribute to adequate daily intakes.

Figure 40 *Snack food is sweet and fast. Unfortunately many young athletes consume snack food daily for pleasure*

Table 7   *Recommended dietary allowances for vitamins*

Vitamin	Age: males			Age: females		
	NRC 15–18 DGE 15–18	19–24 19–25	25–50 25–51	15–18 15–18	19–24 19–25	25–50 25–51
Vit B1 (mg)	1.5/1.6	1.5/1.4	1.5/1.3	1.1/1.3	1.1/1.2	1.1/1.1
Vit B2 (mg)	1.8/1.8	1.7/1.7	1.7/1.7	1.3/1.7	1.3/1.5	1.3/1.5
Niacin (mg)	20/20	19/18	19/18	15/16	15/15	15/15
Vit B6 (mg)	2.0/2.1	2.0/1.8	2.0/1.8	1.5/1.8	1.6/1.8	1.6/1.6
Folate (μg)	200/300	200/300	200/300	180/150	180/150	180/150
Vit B12 (μg)	2/3	2/3	2/3	2/3	2/3	2/3
Vit C (mg)	60/75	60/75	60/75	60/75	60/75	60/75
Vit A (μg) RE	1000/1000	1000/1000	1000/1000	800/900	800/800	800/800
Vit D (μg)	10/5	10/5	5/5	10/5	10/5	5/5
Vit E (mg) TE	10/12	10/12	10/12	8/12	8/12	8/12
Vit K (μg)	65/70	70/70	80/80	55/60	60/60	65/65
Pantothenic acid	[a]/8	[a]/8	[a]/8	[a]/8	[a]/8	[a]/8

The data shown are derived from NRC/DGE.
NRC = National Research Council, Recommended Dietary Allowances, 1989 (USA).
DGE = Deutsche Gesellschaft für Ernährung, Empfehlungen für die Nährstoffzufuhr, 1991.
[a]No allowance given.

In view of concern among athletes to assure adequate intakes in any way and the risks of oversupplementation on the one hand, as well as the invasive/complex nature of determination of vitamin status on the other, athletes and their counselors should be educated about dietary practices and safe daily intakes. Preparations or foods with added vitamins for athletes should be regulated accordingly.

Daily intake of *a low dose* vitamin preparation or nutrient preparations, supplying not more than the recommended daily/safe intake (Table 7), in addition to the normal diet, is advisable in periods of intensive training or in any situation where athletes abstain from a normal diet such as during periods of limited food intake combined with intensive training (especially in females, in vegetarian athletes, and in weight class sports participants, see Table 1, p. 8).

The use of mega doses should be discouraged, especially because of undesired side effects (3) and because of possible negative interactions with other micronutrients (109). Additional research is needed on the effect of higher dosages of anti-oxidant vitamins on athletes' health and recovery processes.

Although the use of mega doses of vitamins by athletes is often defended with the argument that substantial amounts of vitamins may be lost with sweat and urine, there are no scientific data supporting this. Sweat vitamin losses are in general negligible (13, 28, 84).

5
Nutritional Ergogenics

Since the control of anabolic steroids and other illegal ergogenic aids has been intensified, a variety of nutrients have been put forward as effective, safe, and legal alternatives. In the following paragraphs some potential nutritional alternatives to illegal pharmaceuticals, as well as other nutritional supplements marketed for athletes, will be discussed.

SINGLE AMINO ACIDS

Recently various amino acid supplements have been suggested to improve performance, mainly because of their influence on hormone secretion or effects on brain metabolism, concentration and arousal.

Although most amino acid supplements are targeted for strength athletes and bodybuilders, several are also designed for the endurance athlete. Research on the performance enhancing effects of such supplements is increasing, but the available data are limited. The following discussion is based upon several recent reviews (33, 94, 201, 202) and investigations.

Arginine and ornithine

It has been hypothesized that the ingestion of arginine and ornithine may stimulate the release of human growth hormone, which is thought to stimulate muscle growth.

Five recent studies are available on the effect of arginine and/or ornithine supplementation on body composition and/or muscular strength or power. Three of these studies indicated significant increases in lean body mass, an indication of increase in muscle mass and/or decrease in fat mass. However, Williams (202) criticized the experimental methodology of these studies. His recalculation of the data, using appropriate statistical techniques, revealed no significant differences between the supplemented and the placebo treated groups. The other two studies with a correct methodological approach have so far appeared only as abstracts. Both studies reported no significant effect of arginine or a mixture of different amino acids on measures of strength, power, or growth hormone, in well trained weight lifters (81, 192). Currently there are no sound research data to support an ergogenic effect of arginine and ornithine. This may be related to the general ineffectiveness of the amino acids in increasing growth hormone levels beyond the range in which normal physiological levels daily fluctuate. It is suggested that the very high dosages needed to cause any significant effect may cause gastric distress (33). Additional research is necessary to evaluate the hypothetical value of supplementing these single amino acids in athletes.

Tryptophan and branched chain amino acids (BCAAs)

Tryptophan may also increase growth hormone, but its theoretically most potent ergogenic effect is based upon another mechanism, the formation of 5-hydroxytryptamine and serotonin in the brain. Segura and Ventura (170) suggested that these neurotransmitters may improve performance by increasing the tolerance to pain during intense exercise. In support of their hypothesis, they found that 1200 mg of tryptophan consumed in 300 mg doses over a 24 hour period increased time to exhaustion and reduced a rating of perceived exertion (RPE) on a treadmill run to exhaustion at an exercise intensity of 80% VO_2max. Additional research is needed to confirm this finding.

In contrast to the hypothesis of Ventura, Newsholme (140) has suggested that serotonin may be involved in the development of fatigue because of its depressive activity. Thus, entry of tryptophan into the brain may contribute to development of fatigue. Based upon research with animals, which suggests that low blood levels of branched chain amino acids (BCAAs) may facilitate the entry of tryptophan into the brain, Newsholme theorizes that a decrease in serum levels of BCAAs during the later stage of endurance exercise, which induces a relative increase of the levels of blood tryptophan, could be a contributing factor to fatigue. Theoretically, BCAA supplements, enhancing the blood BCAA concentration and thus decreasing the tryptophan uptake in the brain by competing for the transport system across the blood–brain barrier, could help to delay the onset of fatigue. Unfortunately, few data are available to support this hypothesis. For example, some research has revealed no change in the tryptophan : BCAA ratio at the completion of a 42.2 km marathon. In a direct test of this hypothesis, Vandewalle *et al.* (184) depleted subjects of muscle glycogen and then subjected them to a cycle ergometer ride to exhaustion at 75% VO_2max; they reported no beneficial effect of BCAA supplementation. Nor did Galiano and others (67), who supplemented BCAA during prolonged exercise to exhaustion at a workload of 70% VO_2max. Kreider and his associates (99, 122) provided BCAA supplements to five triathletes for 14 days prior to and during a "half-Ironman" triathlon (2 km swim; 90 km bike; 21 km run) under laboratory conditions. Although no significant differences were noted between the BCAA and placebo conditions, the authors suggested run performance in the last segment might be enhanced. Therefore, additional research with prolonged endurance tasks is necessary to test the validity of Newsholme's hypothesis and the value of BCAA supplementation during exercise.

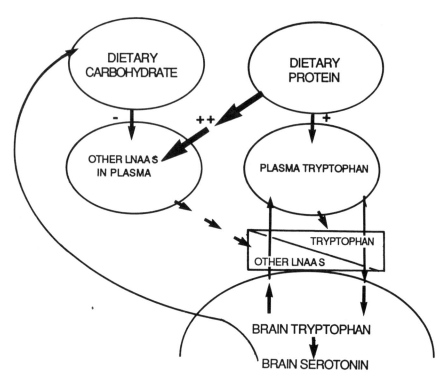

Figure 41   *The entry of the amino acid tryptophan, a precursor for serotonin, depends on the ratio tryptophan/other large neutral amino acids (LNAAs) (leucine, valine, isoleucine) in the blood. Carbohydrate ingestion will, via enhanced insulin secretion, lower the LNAA level in blood. This will increase tryptophan entry into the brain. Protein consumption will enhance plasma LNAAs more than plasma tryptophan. This will reduce tryptophan uptake. When serotonin is high, a feedback will lead to reduced CHO consumption. Low serotonin levels may induce CHO snacking. Adapted from R. J. Wurtman, Nutrients that modify brain function, Scientific American 1982; 246: 42–51. Copyright © by Scientific American, Inc. All rights reserved*

ASPARTATES

The potassium and magnesium salts of aspartate, a nonessential amino acid, have been postulated to improve performance by several mechanisms. The prevailing hypothesis is that aspartates will reduce the accumulation of blood ammonia during exercise. Increases in blood serum ammonia have been correlated with muscular and central fatigue (8, 29, 189).

Research data regarding the stimulating effect of aspartates are equivocal. A number of studies have reported no effect. As an example of a well designed study, Maughan and Sadler (116) gave a placebo or 3 g each of potassium and magnesium aspartame to eight subjects 24 hours prior to a cycle ergometer ride to exhaustion and reported no beneficial effects. Conversely, an equal number of other studies have documented a positive effect on performance, some reporting greater than 20% improvement in aerobic endurance. For example, Wesson and others (197), who used an appropriate research design, gave a placebo or 10 g of aspartates to subjects over a one day period prior to exercise to exhaustion at 75% VO$_2$max. They reported a significant decrease in serum ammonia levels and a 15% increase in endurance performance. It is clear that additional research is needed, particularly with dosages of 10 g or more, for such high dosages have often been associated with enhanced performance. No toxic effects have been reported in studies using these dosages.

L-CARNITINE

Carnitine is a vitamin-like compound that primarily facilitates the transport of long chain fatty acids into the mitochondria for use in energy production. It has often been suggested that increased carnitine availability may increase the use of fat as substrate for energy production and that this could lead to a sparing of muscle glycogen during exercise. This might increase the time to exhaustion. However, the available data are not supportive of this viewpoint.

Data from early studies were inconsistent due to inadequate research design or dosage utilized. For example, although several studies from Otto's research group (146, 175) found no effect of 500 mg carnitine taken daily for 4 months on free fatty acid utilization, VO$_2$max, anaerobic threshold, exercise time to exhaustion, or work output on a cycle ergometer for 60 minutes, Bucci (34) criticized these reports because of low dosage. However, three more recent studies, using doses up to 2 g, also reported no effects of carnitine supplementation on

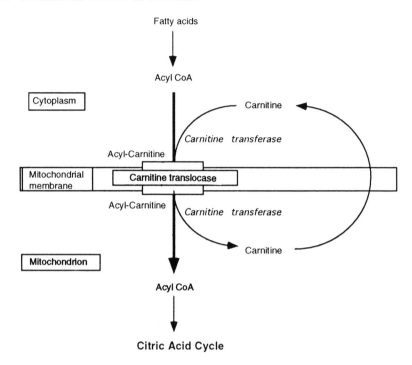

Figure 42   *Carnitine is needed to transport long chain fatty acids across the mitochondrial membrane. Activated fatty acids are linked to carnitine through formation of acyl carnitine by the enzyme carnitine transferase. This compound passes the membrane, after which the process is reversed and carnitine is transported back to the cytoplasm. The fatty acid will enter the citric acid cycle to be oxidized. Adapted from R. Laschi, L-carnitine and Ischemia, Foundazione Sigma-Tau, 1987*

fuel utilization at 50% VO_2max, maximal heart rate, anaerobic threshold, VO_2max, or exercise time to exhaustion (73, 147, 208). Recently Wagenmakers (190) reviewed the possible effects of carnitine and concluded that oral L-carnitine does not affect endurance performance or fat metabolism. D,L-Carnitine has been shown to be harmful and should not be taken. L-carnitine, which is produced in the human body, is relatively harmless when taken orally. However, oral supplementation in healthy human subjects does not lead to increased levels of L-carnitine in muscle, and thus fails to affect muscle energy/fat metabolism in healthy, trained individuals (190); nor does exercise lead to a decrease in muscle carnitine content (50, 95).

CoQ10 (UBIQUINONE)

CoQ10 is a lipid present in the mitochondria, particularly in the heart. It has been used therapeutically for the treatment of cardiovascular disease because of its role in oxidative metabolism and as an antioxidant. Because CoQ10 has increased oxygen uptake and exercise performance in cardiac patients, it has been suggested that it may be effective for endurance athletes as well. However, there are no data available to support this.

Several recent studies found that although CoQ10 supplementation may significantly increase serum CoQ10 levels, compared to a placebo, there were no significant improvements in serum glucose or lactate at submaximal or maximal workloads, cardiovascular functions such as heart rate, VO_2max, or endurance performance (21, 159, 210).

Additional research with CoQ10 is needed. However, Demopoulos and others (51) did suggest that supplementation with this compound may actually be hazardous, for it may induce free radical formation, which could lead to cell membrane lipid peroxidation and damage.

INOSINE

Inosine is a nucleoside. Some of the reported metabolic roles of inosine such as facilitation of ATP (energy rich phosphate) synthesis, effects on muscle glycogen breakdown, and on blood and oxygen supply have been extrapolated to exercise physiology, suggesting that both strength and endurance athletes might benefit from supplementation. Inosine is available either pure or combined with other cofactors, such as CoQ10. Only one study has investigated its effect on endurance parameters. Williams and others (200) used a recommended supplementation protocol (6 g of inosine for 2 days) and reported no significant effects on metabolic parameters and performance. Clearly additional research is needed to evaluate any hypothetic benefit.

BEE POLLEN

Chemical analysis of bee pollen shows that these are composed of a mixture of vitamins, minerals, amino acids, and other nutrients. Although bee pollen does not appear to induce any specific physiologic effect, its theoretical ergogenic effect may be based on the roles that vitamins and minerals are thought to have in exercise metabolism. To test this, highly trained runners were studied, and no significant effect on the rate of recovery, as measured by performance in repeated maximal treadmill runs to exhaustion with set recovery periods, could be found (205). Additional well-controlled studies have reported no effects on maximal oxygen consumption or other physiological responses to exercise, nor on endurance performance (36, 179). Individuals who are allergic to bee pollen may experience an anaphylactic reaction (54).

PHOSPHATE SALTS

Phosphorus is an essential nutrient that functions in the body as phosphate salt. This is a cofactor or component for several B vitamins, ATP and phosphocreatine, 2,3-DPG (diphosphoglycerate), and an intracellular buffering system. Based on these metabolic roles, phosphate supplements have been suggested to have performance-improving characteristics. Some early studies suggested that phosphate salt supplementation was an effective ergogenic for several types of physical performance. Although Boje (20) criticized these studies for design flaws, he indicated that phosphates probably could increase physical performance if consumed in quantities found in the normal diet. Most of the current research, however, has focused on the ability of phosphate salts to enhance oxygen uptake and endurance performance. Although some studies support Boje's observation over 50 years ago, the data are still equivocal (100, 202).

Williams (202) has cited four studies, all using appropriate experimental designs and dosages, reporting no beneficial effect of phosphate supplementation on such physical performance parameters as cardiovascular function and oxygen efficiency at 60% VO_2max, lactic acid production, VO_2max, or performance in an 8 km (5 mile) bike race. In contrast, four other well-designed studies reported significant increases in VO_2max approximating 10%, decreases in lactate production during submaximal exercise, enhancement of myocardial efficiency, increases in running and cycling time, and decreases in the time to complete a 40 km bicycle race under laboratory conditions. The dosage in these studies was approximately 4 g daily of sodium phosphate consumed, usually consumed in 1 g doses, for 3–6 days.

Thus, additional research is needed to confirm the reported ergogenic effects.

SODIUM BICARBONATE

Sodium bicarbonate is an alkaline salt. Its major function is to control acid–base balance in the blood and its proposed role as an ergogenic substance is to buffer the lactic acid which is produced during high-intensity exercise and accumulates in the blood. Such increased buffering may affect the onset of fatigue. Research on the ergogenic effects of sodium bicarbonate has been conducted for over 50 years and its effectiveness is still debatable.

In many studies a quantity 0.15–0.40 g (most often 0.30 g) per kilogram body weight was administered 1–3 hours prior to an exercise task of maximal intensity and short duration. Such a performance requires mainly muscle glycogen as energy source and results in the production of lactic acid, which is supposed to induce fatigue. Usually these tests were performed to exhaustion and consisted of single bouts of exercise or repeated bouts of sprint exercise with small rest periods in between.

About 50% of the studies of acceptable quality have shown a beneficial effect on physical performance and on psychological perceptions of exertion. Several major reviews regarding the effectiveness of sodium bicarbonate have been published (70, 117, 85). In one extensive review (85), it was hypothesized that sodium bicarbonate supplementation with an appropriate dosage appears to have no effect on high-intensity performance of 30 seconds or less, nor on endurance performance that depends primarily upon oxidative metabolism, but will enhance performance in intense continuous exercise of approximately 1–7.5 minutes or in repetitive bouts of intense exercise involving short rest intervals.

Several studies have reported gastrointestinal distress, such as diarrhea, following ingestion of sodium bicarbonate, while several case studies of gastric wall damage have also been reported.

No gastrointestinal problems were reported with sodium citrate, which has the same effects on buffering capacity, in dosages up to 0.5 g/kg body weight (120).

6
Summary

Adequate intake of nutrients

In athletes, adequate intake of nutrients is of essential importance for the maintenance of an appropriate nutritional status, optimal performance, adequate recovery and the reduction of health risks. Extremely high energy turnovers, e.g. as in intensive endurance exercise, require adequate energy and nutrient intakes, to maintain energy, nitrogen and fluid balance.

The large gastrointestinal bulk associated with a carbohydrate rich diet, when consuming normal food, causes athletes to change their food habits and to ingest 30–50% of daily energy intake as in-between meals/snacks, often high in energy but low in dietary fiber, protein and micronutrients. This leads to a reduction in diet quality, unless food products and/or dietary supplements are chosen which are of appropriate composition. Education of athletes and coaches is important in this respect. However, although many nutrition oriented publications appear in athletic journals, the evidence shows that nutritional knowledge remains poor (see chapters 1, 2 and 4).

Athletes are known to have an increased exercise-induced utilization/loss of macro- and micronutrients. This loss should normally be compensated by the daily diet. However, the nutrient density of protein and most micronutrients in normal food is such that these often become inadequate. Therefore, athletes who ingest chronically or repeatedly low energetic diets, such as gymnasts, dancers, low weight class athletes, bodybuilders and female distance runners, are at potential risk of marginal nutrition. Such athletes can improve their nutrient intake and nutrition status by choosing normal foods, food products and supplements which have an enhanced nutrient density for specific nutrients (see chapters 1 and 4).

Carbohydrate

Carbohydrate (CHO) is the most important nutrient for high intensity performance. Energy release from CHO is up to three times as fast as from fat. However, CHO stores in the body are small, which limits the time to perform high intensity exercise. Apart from decreasing performance, CHO depletion induces an increased utilization of protein for energy production. This results in the production of ammonia, which may enhance fatigue. CHO ingestion during exercise allows sparing of the body's CHO stores, reduction of protein utilization and ammonia production, and a delay of fatigue/improvement of performance. Adequate CHO ingestion between training sessions/days or intense performance is of utmost importance to avoid progressive fatigue development/overtraining. CHO sources used during exercise should be rapidly absorbable, i.e. have a high glycemic index, and should be combined with sufficient fluid intake (chapter 2).

Fat

Fat is a "slow" energy source. When using fat as prime energy source, athletes can only work at 40–60% of their maximal capacity. Nevertheless, increased fat utilization, as a result of training, reduces the use of CHO from the stores in the body, and thus will influence CHO availability and fatigue. Daily fat intake in athletes should be relatively low, <30 energy%, allowing for an increase in the proportion of CHO in the diet.

Saturated fat sources should be avoided and vegetable-, fish- and plant-oil "based foods" should be promoted (chapter 2).

Protein

Protein is needed for muscle growth, repair of tissues and enzymatic adaptation. The building blocks of protein, amino acids, are involved in numerous metabolic pathways and processes. Some of the amino acids are known to influence hormone production and neurotransmission. The latter is speculated to influence fatigue/performance.

The protein requirement of athletes is increased and, according to present knowledge, amounts to approximately 1.2–1.8 g/kg body weight. The reason for this increase is enhanced utilization of amino acids in oxidative energy production during physical exercise, a process which is known to be intensified at higher work levels and in a state of carbohydrate store depletion. Athletes who ingest low caloric diets will have low protein intakes, which may not compensate for the net nitrogen loss from the body and will influence synthesis processes and training adaptations. To these categories belong bodybuilders, weight class athletes, gymnasts, dancers and female long distance runners, and under some circumstances vegetarian athletes. Protein intake/supplementation above levels normally required will not enhance muscle growth or performance. The use of single amino acids, to influence metabolic pathways involved in fatigue development and hormone production, needs further research to make definite statements (chapters 2 and 5).

Fluid and electrolytes

Fluid and electrolytes are of prime concern during prolonged physical exercise, especially in the heat. Progressive fluid loss from the body, by means of sweating and breathing, and in endurance events also by diarrhea, is associated with a decreased blood flow through the extremities, a reduced plasma volume and central blood volume, a reduction in sweating and heat dissipation, and under circumstances of high-intensity work in the heat with heat stroke/collapse. Dehydration of >1.5 liters is known to reduce the oxygen transport capacity of the body and to induce fatigue. Appropriate rehydration is known to counter these effects and to delay fatigue. In contrast to plain water, addition of sodium and CHO (up to 80 g/l) to rehydration drinks is known to stimulate water absorption, to influence water balance regulating hormones less dramatically and to supply energy. Addition of other electrolytes should not exceed the levels of loss with whole body sweat. Sport rehydration drinks should not be hypertonic (chapter 3).

Minerals

Minerals are important substances for the musculoskeletal system, as well as for numerous biological actions. Exercise is known to be associated with increased mineral losses, through sweating, during exercise, and through urine in the post exercise phase. As with most nutrients, mineral intake depends on the quality of the diet and the amount of energy consumed. Therefore, athletes consuming low energetic diets are at risk of marginal mineral intake, especially of magnesium. Vegetarian athletes are especially prone to iron deficiency.

Athletes may develop an impaired mineral status if the quality of the diet is poor. Impaired iron, zinc and magnesium status are known to induce malperformance and muscle weakness and are often associated with muscle cramp. The latter needs further research to validate the direct influence of mineral deficits (chapters 1 and 4).

Trace elements

The role of trace elements for athletes has only received attention in recent years. As with minerals, trace elements are increasingly lost as a result of intensive physical training. Trace element losses with sweat (copper) and urine (chromium) may exceed the daily recommended intakes. The composition of the diet may affect these losses. For example, high CHO intakes, especially of high glycemic index carbohydrates, are known to enhance losses of chromium, whereas diets rich in dietary fiber, often consumed by endurance athletes and vegetarians, are known to reduce trace element absorption. Athletes consuming low energetic diets are at risk of marginal trace element intake. Now that it is recognized that exercise leads to enhanced tissue/cell damage, the importance of selenium, which acts within the free radical scavenging processes, has received attention. Much research is needed in this field, but it is felt that supplementation with amounts not exceeding the recommended safe daily intakes will contribute to adequate daily intake in athletes (chapter 4).

Vitamins

Vitamins have received widespread attention. They are essential cofactors in many enzymatic reactions involved in energy production and in protein metabolism. Any shortage of a vitamin is therefore linked to suboptimal metabolism, which in the long term will result in decreased performance or even illness. In addition, some vitamins act as antioxidant substances and are believed to have a protective role for tissue/cell integrity, which in the case of metabolic stress may be threatened.

Vitamin supplementation has been shown to restore performance capacity in cases of vitamin deficit and to reduce tissue damage due to free radicals. Vitamin supplementation with quantities exceeding those needed for optimal/blood levels has not been shown to improve performance. As with minerals and trace elements athletes involved in intensive training, but consuming low energetic diets, are the most prone to marginal vitamin intakes. In general it can be concluded that vitamin restoration of energy dense processed foods or supplementation with preparations will not enhance performance but may, in athletic populations, contribute to adequate daily intakes.

Daily intake of a low dose vitamin preparation or nutrient preparations, supplying not more than the recommended daily/safe intake, may be advisable in periods of intensive training or in any situation where athletes abstain from a normal diet such as during periods of limited food intake combined with intensive training (especially in females and in weight class sports participants) (Table 1, chapters 1 and 4).

Nutritional ergogenics

There are many substances which are part of our daily nutrition. Substances such as caffeine, carnitine, aspartates, sodium bicarbonate, bee pollen, specific amino acids, etc., have recently received scientific attention due to their possible influence on performance, fatigue and recovery. In many cases such substances need further research to produce conclusive scientific evidence. Some of these substances have been described in Chapter 5.

The future will bring more and more knowledge about, and possibilities for, nutritional intervention and supplementation. This development will be stimulated by increasing knowledge about nutritional substances involved in the different metabolic pathways, including brain metabolism. Judgement about the role of food derived substances, as being nutritional or pharmacological, requires research and optimal interaction between science and the food industry.

7
A Brief Outline of Metabolism

GLYCOGEN

Glycogen is a glucose polymer. It is a storage form of glucose in human muscle and liver, comparable to the storage of glucose in plant starches such as potato and banana. Glycogen is synthesized or broken down by different enzymes, within the cytoplasm. When synthesized, glucose is phosphorylated to glucose-1-phosphate. Glucose-1-phosphate is converted to uridine diphosphate (UDP) glucose, which is built into glycogen by the action of the enzyme glycogen synthetase. When the amount of glucose is insufficient, glycogen is broken down by the enzyme glycogen phosphorylase. Glycogen is mainly synthesized in periods when the amount of glucose present in cells exceeds the amount which is needed for energy production. Glycogen metabolism in the liver regulates the blood glucose level. After meals, glucose and fructose are taken up by the liver, leading to a storage of liver glycogen. During the night or during fasting, liver glycogen will be broken down to maintain a normal blood glucose level. Muscle glycogen is primarily meant to be a rapid energy source, to be available in a situation of sudden intensive muscular work.

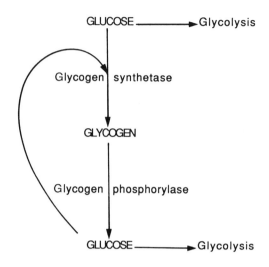

GLYCOGEN METABOLISM

Synthesis or degradation of glycogen in the liver and the muscle is regulated by many factors.

Synthesis will normally take place if the supply of glucose "building units" exceeds the need of glucose for energy production, i.e. if the amount of glucose within the cell becomes increased.

This situation occurs after meals, when during a state of physical relaxation the digestion and absorption of carbohydrate leads to increased blood glucose levels in a hormonal milieu which favors synthesis. Thus, insulin will be high, glucagon and stress hormones will be low. In this situation the cells will take up glucose and the enzyme glycogen synthetase will be activated (+), whereas glycogen phosphorylase will be inhibited (−).

In the case of a rapid energy requirement, a number of signals of central nervous system and hormonal origin will cause stress hormones and glucagon to be increased and insulin to be decreased. The enzyme glycogen synthetase will be inhibited (−) and the degrading enzyme glycogen phosphorylase will

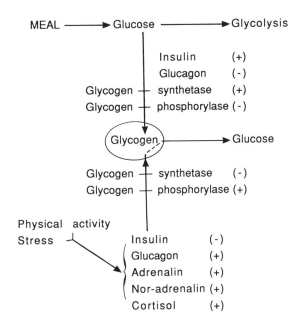

be activated (+), resulting in the liberation of glucose-1-phosphate from the glycogen pool.

GLUCOSE METABOLISM

When used in energy production, glucose enters the glycolytic pathway in which it is converted in a number of steps to pyruvate.

Depending on the quantitative need for energy, pyruvate is either largely converted to lactic acid—this is the case during intensive stimulation of glycolysis, such as during supra-maximal sports activities (0.5–3 minutes duration)—or is taken up in the oxidative energy pathway, the citric acid (Krebs) cycle—mainly during endurance events.

The pathway glucose → lactic acid is reversible, which means that a high lactic acid content in the blood after intensive sports activity can be lowered by conversion of lactate via a different way called gluconeogenesis, a partial reversal of glycolysis back

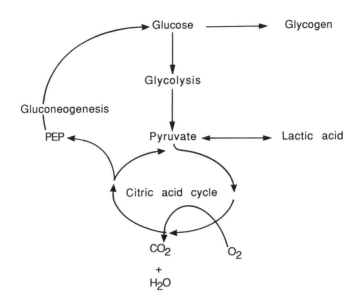

to glucose, which can be stored again as glycogen. Lactate can also be oxidized or converted to fat.

During the conversion of glucose to lactate, 2 moles of ATP are produced per mole of glucose. During complete oxidation of glucose within the citric acid cycle, pyruvate is converted to water and carbon dioxide and a total of 36 moles of ATP are produced.

ADIPOSE TISSUE/TRIACYLGLYCEROL

Fatty acids are stored in the body as triacylglycerols (triglyceride) in fat cells which make up the adipose tissue. Fat is also stored in muscle tissue in the form of triglyceride, present in small intramuscular fat droplets.

After a meal, fat is absorbed and circulates in the blood as triglycerides in the form of circulating lipid particles (high density lipoprotein (HDL), very low density lipoprotein (VLDL), low density lipoprotein (LDL), chylomicrons) or as

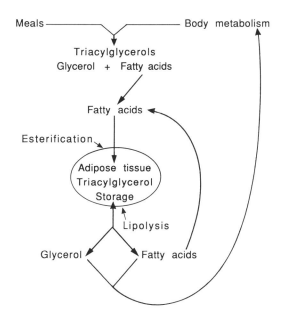

free fatty acids bound to albumin, called non-esterified fatty acids (NEFAs). As with glycogen, the synthesis of fat or its degradation depends on the concentration of the "building blocks", in this case fatty acids. This concentration is determined mainly by uptake or release of free fatty acids in and from triacylglycerols and their withdrawal for energy metabolism.

Thus, when energy production is low, the supply of fatty acids after a meal will lead to an increase in the fatty acid concentration within the cell. This will stimulate esterification and the amount of triacylglycerol within the fat cell will increase. Such a process is mediated by a large number of interactions, in which hormonal and nervous influences play a major role. In the case of increased energy requirement, fatty acids will be used in energy production. This will result in a decrease in the fatty acid concentrations, which will stimulate the breakdown of triacylglycerols into glycerol and free fatty acids to compensate for this.

TRIACYLGLYCEROL METABOLISM

The process of binding fatty acids (esterification) in the form of triglyceride and their release from it is called the triglyceride/fatty acid cycle.

The activity of this cycle is determined by the metabolic need for fatty acids for energy production and by the supply of fatty acids from external sources.

The glycerol necessary for esterification is derived from glycolysis.

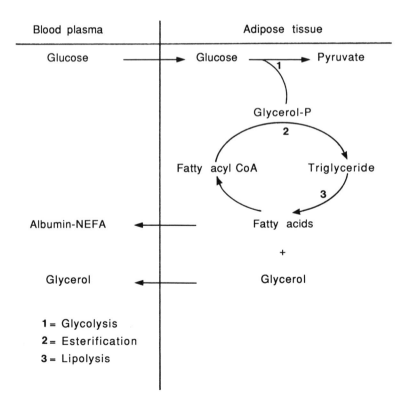

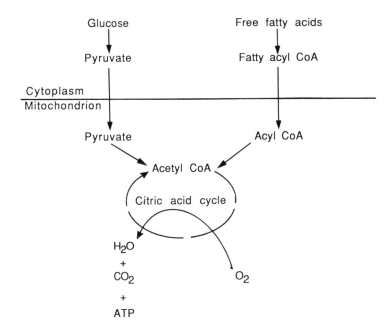

FATTY ACID METABOLISM

Free fatty acids are metabolized by aerobic metabolism within the citric acid cycle.

For this chain of metabolic steps, fatty acids are converted to fatty acyl CoA which can enter the citric acid (Krebs) cycle where it is converted to acetyl CoA. At high rates of fat oxidation there is a high production of acetyl CoA and citrate, the first citric acid cycle intermediate formed from acetyl CoA. Acetyl CoA is known to inhibit the conversion of pyruvate to acetyl CoA. Additionally, citrate will inhibit glycolysis. Thus, increased fatty acid oxidation inhibits both the rate of glycolysis and the first conversion step of pyruvate in the citric acid cycle. As a result, total carbohydrate oxidation will be reduced.

Conversely, increased carbohydrate metabolism, e.g. after intake of oral CHO, inhibits lipolysis, reduces the availability of fatty acids and thus their oxidation. In exercise metabolism these processes of carbohydrate and fat utilization are tightly coupled and controlled by nervous and hormonal mechanisms. They may be influenced by exogenous supply of either carbohydrate or fat, or by substances which stimulate the metabolism of either substrate.

PROTEIN

All protein in the body is *functional* protein.

We do not have a protein store as is the case with carbohydrate in the form of glycogen or fat, stored as triacylglycerol in adipose tissue.

The amount of functional protein depends on organ function. Increased functioning, e.g. work intensity of heart or skeletal muscle, will result in a stimulus to build up more contractile protein. As result, the muscle will hypertrophy. Increased metabolic demand will lead to an increased number of enzymes and mitochondria etc.

Amino acids are the building blocks of protein. Essential amino acids cannot be produced by the body. Therefore, we need appropriate protein sources to supply these amino acids. Periods of growth are characterized by increased protein synthesis, periods of illness or inactivity by increased protein degradation. In both instances the amount of amino acids and nitrogen needed is increased. Appropriate daily protein intake is therefore the key in the maintenance of nitrogen balance.

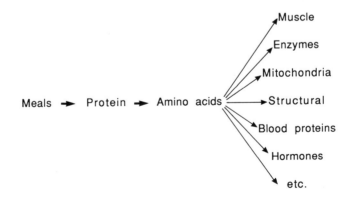

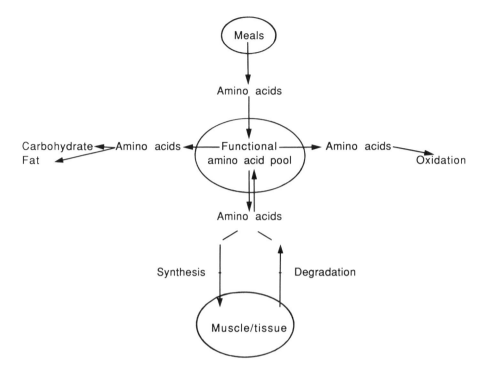

PROTEIN METABOLISM

The substances required for protein synthesis or resulting from protein breakdown are amino acids. Amino acids form a functional amino acid pool in blood and in tissue fluids.

Protein which is broken down, i.e. protein supplied with meals or protein within the body itself, results in a supply of amino acids to this pool. With an appropriate supply of amino acids, shortly after a meal, protein synthesis may be enhanced due to the combination of high insulin and appropriate amino acid supply. The amino acids which are not used in protein synthesis will either be oxidized or converted to carbohydrate and fat.

A result of these processes is that the concentration of most amino acids in blood and tissue fluids is kept within a narrow range.

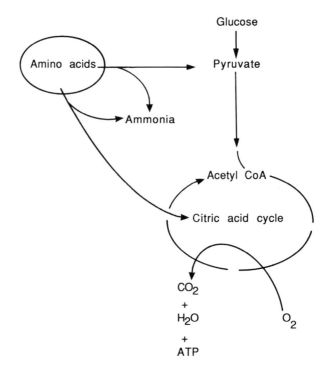

AMINO ACID OXIDATION

The general strategy of amino acid degradation is to produce major metabolic intermediates that can be converted into glucose and fat or can be oxidized by the citric acid (Krebs) cycle. Most of the amino acids are oxidized within the liver and some of them—the branched chain amino acids—also in the muscle.

Amino acid oxidation takes place in the mitochondria and is always increased in periods of physical exercise. This increased oxidation is primarily the result of a change in the anabolic/catabolic hormonal milieu towards catabolism.

Amino acid oxidation is further enhanced when the carbohydrate pools in the body become depleted. Available evidence suggests that this results in an increased amino acid requirement of 1.2–1.8 g/kg body weight per day in daily training endurance athletes.

ENERGY METABOLISM

During the initial stage of sudden physical exercise, the extra energy needed is mainly produced by the breakdown of muscle glycogen to lactate. Blood glucose does not contribute substantially during the first minutes of exercise. At this stage the

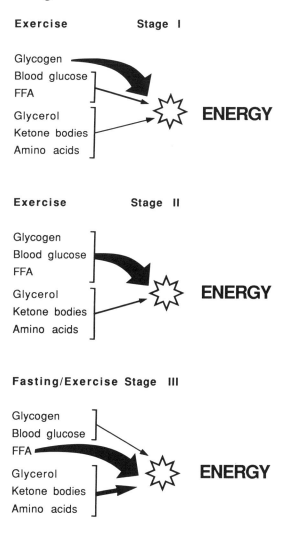

The size of the arrow indicates roughly the quantitative contibution to energy production

glycogenolysis of the liver has to be increased. The formed lactate is released into the bloodstream and taken up by the liver, the heart and non-active muscle tissue, where it is either oxidized or resynthesized to glucose. At a later stage as glucose production from the liver is significant, muscle will increasingly use blood glucose for energy production.

Additionally, lipolysis in fat cells—initially a gradually increasing process—has led to high blood fatty acid levels, through which the contribution of fatty acids for energy production increases. Fatty acids become more and more oxidized in muscle and liver. Ketone bodies, which result from incomplete fat oxidation in the liver, are taken up from blood by the heart and the muscle for their final oxidation.

With increasing metabolic stress, especially in conditions of carbohydrate depletion, synthesis of protein may be decreased and the degradation of amino acids increases.

Degradation of amino acids in muscle and liver finally leads to the production of urea which will be excreted with urine and sweat. The carbon skeletons of the amino acids will enter the citric acid cycle in liver, where they will be used for gluconeogenesis, and muscle, where they will be oxidized.

With ongoing exercise and also during fasting, the endogenous carbohydrate stores in liver and muscle will become depleted. If no glucose were to be produced from gluconeogenic precursors in liver and kidney, the blood glucose level would drop sharply. Gluconeogenic precursors are amino acids, glycerol and lactate. At the same time, fat oxidation will be maximized, resulting in a reduced need for carbohydrate. Ketone bodies resulting from fat metabolism in the liver will be metabolized by heart, muscle and with prolonged fasting also in the brain. Under these circumstances, maximal work capacity will drop to approximately 50%, due to the lack of carbohydrate.

References

(1) Ahlborg, G., Felig, P., Hagenfeldt, L. *et al*. Substrate turnover during prolonged exercise in man. *J Clin Lab Invest* 1974; 53: 1080–1090.

(2) Alhadeff, L., Gualtieri, T., Lipton M. Toxic effects of water-soluble vitamins. *Nutr Rev* 1984; 2: 33–40.

(3) Anderson, R. A., Polansky, M. M., Bryden, N. A. *et al*. Strenuous exercise may increase dietary needs for chromium and zinc. In Katch, F. I. (Ed.), *Sport, Health and Nutrition*. 1984 Olympic Sci Cong Proc, Vol. 2. Human Kinetics Publishers, Champaign, Illinois, 1986, pp. 83–88.

(4) Anderson, R. A., Polanski, M. M., Bryden, A. *et al*. Trace minerals and exercise. In Terjung, R., Horton, E. S. (Eds), *Exercise, Nutrition and Energy Metabolism*. Macmillan, New York, 1988, pp. 180–195.

(5) Anderson R. A. New insights on the trace elements, chromium, copper and zinc, and exercise. In Brouns, F., Saris, W. H. M., Newsholme, E. A. (Eds), *Advances in Nutrition and Top Sport. Med Sport Sci*, Vol. 32. Karger, Basel, 1991, pp. 38–58.

(6) Armstrong R. B. Muscle damage and endurance events. *Sports Med* 1986; 3: 370–381.

(7) Auclair, E., Sabatin, P., Servan, E., Guezennec, C. Y. Metabolic effects of glucose, medium chain triglyceride and long chain triglyceride feeding before prolonged exercise in rats. *Eur J Appl Physiol* 1988; 57: 126–131.

(8) Banister E. W., Cameron B. J. Exercise-induced hyper-ammonemia: peripheral and central effects. *Int J Sports Med*, May 1990, 11 (Suppl 2): S129–S142.

(9) Barr, S. I. Women, nutrition and exercise: a review of athletes' intakes and a discussion of energy balance in active women. *Prog Food Nutr Sci* 1987; 11: 307–361.

(10) Beek van der E. J., Dokkum, W. van, Schrijver, J. Marginal vitamin intake and physical performance in man. *Int J Sports Med* 1984; 5: 28–31.

(11) Beek van der E. J. Vitamins and endurance training: Food for running or faddish claims? *Sports Med* 1985; 2: 175–197.

(12) Beek E. J. van der, Dokkum W. van, Schrijver, J. Controlled vitamin C restriction and physical performance in volunteers. *J Am Coll Nutr* 1990; 9(4): 332–339.

(13) Beek E. J. van der, Vitamin supplementation and physical exercise performance. *J Sports Sci* 1991; 9: Special Issue, pp. 77–89.

(14) Bendich, A. Exercise and free radicals: effects of antioxidant vitamins. In Brouns, F., Saris, W. H. M., Newsholme, E. A. (Eds), *Advances in Nutrition and Top Sport*. Vol. 32. Karger, Basel 1991, pp. 59–78.

(15) Bergström, J., Hultman E. The effect of exercise on muscle glycogen and electrolytes in normals. *Scand J Clin Lab Invest* 1966; 18: 16–20.

(16) Bergström, J., Hultman E. A study of the glycogen metabolism during exercise in man. *Scand J Clin Lab Invest* 1967; 19: 218–228.

(17) Bergström, J., Hultman E. Synthesis of muscle glycogen in man after glucose and fructose infusion. *Acta Med Scand* 1967; 182: 93–107.

(18) Bergström, J., Fürst, P., Holström, B. *et al*. Influence of injury and nutrition on muscle water and electrolytes. *Ann Surg* 1981; 193: 810–816.

(19) Bjorntorp P. Importance of fat as a support nutrient for energy: metabolism of athletes. *J Sports Sci* 1991; 9: Special Issue, pp. 71–76.

(20) Boje, O. Doping: A study of the means employed to raise the level of performance in sport. *League of Nations Bulletin of the Health Organization* 1939; 8: 439–469.

(21) Braun, B., Clarkson, P., Freedson, P. *et al*. The effect of coenzyme Q10 supplementation on exercise performance, VO_2max, and lipid peroxidation in trained cyclists. *Int J Sport Nutr 1*, 1991 (in press).

(22) Bremer, J., Osmundsen, H. Fatty acid oxidation and its regulation. In S. Numa (Ed.), *Fatty Acid Metabolism and its Regulation*. Elsevier, Amsterdam, 1984, pp. 113–154.

(23) Brouns, F., Saris W. H. M., ten Hoor, F. Dietary problems in the case of strenuous exercise: part 1—a literature review, Part 2—the athletes diet: nutrient dense enough? *J Sports Med Phys Fitness* 1986; 26: 306–319.

(24) Brouns, F., Saris, W. H. M., Rehrer, N. J. Abdominal complaints and gastrointestinal function during long-lasting exercise. *Int J Sports Med* 1987; 8: 175–189.

(25) Brouns, F., Saris, W. H. M., Beckers, E. *et al.* Metabolic changes induced by sustained exhaustive cycling and diet manipulation. *Int J Sports Med* 1989; 10 (Suppl 1): S49–S62.

(26) Brouns, F., Rehrer, N. J., Saris, W. H. M., Beckers, E., Menheere, P., ten Hoor, F. Effect of carbohydrate intake during warming-up on the regulation of blood glucose during exercise. *Int J Sports Med* 1989; 10 (Suppl 1): S68–S75.

(27) Brouns, F., Rehrer, N. J., Beckers, E., Saris, W. H. M., Menheere, P., ten Hoor, F. Reaktive Hypoglykämie. *Deutsche Zeitschr Sportmed* 1991; 42(5): 188–200.

(28) Brouns, F., Saris, W. H. M. How vitamins affect performance. *J Sports Med Phys Fitness* 1989 29(4): 400–404.

(29) Brouns, F., Beckers, E., Wagenmakers, A. J. M., Saris, W. H. M. Ammonia accumulation during highly intensive long-lasting cycling: individual observations. *Int J Sports Med* 1990; 11 (Suppl 2): S78–S84.

(30) Brouns, F. Gastrointestinal symptoms in athletes: physiological and nutritional aspects. In Brouns, F., Saris, W. H. M., Newsholme, E. A. (Eds), *Advances in Nutrition and Top Sport. Med Sport Sci*, Vol. 32. Karger, Basel, 1991, 166–199.

(31) Brouns, F. Etiology of gastrointestinal disturbances during endurance events. *Scand J Med Sci Sports* 1991; 1: 66–77.

(32) Brouns, F. Heat—sweat—dehydration—rehydration: a praxis oriented approach. *J Sports Sci* 1991; 9: Special Issue, pp. 143–152.

(33) Bucci, L., Hickson, J., Pivarnik, J., *et al.* Growth hormone release in bodybuilders after oral ornithine administration. *FASEB J* 1990; 4: A397 (abstract).

(34) Bucci, L. Nutritional erogenic aids. In Hickson, J., Wolinsky, I. (Eds), CRC Press, Boca Raton, Florida, 1989.

(35) Campbell, W. W., Anderson, R. A. Effects of aerobic exercise and training on the trace minerals, chromium, zinc and copper. *Sports Med* 1987; 4: 9–18.

(36) Chandler, J., Hawkins J. The effect of bee pollen on physiological performance. *Int J Biosoc Res* 1984; 6: 107–114.

(37) Clarkson, P. M. Minerals: exercise performance and supplementation in athletes. *J Sports Sci* 1991; 9: Special Issue, pp. 91–116.

(38) Clement, D. B., Sawchuk, I. L. Iron status and sports performance. *Sports Med* 1984; 1: 65–74.

(39) Costill D. L. Carbohydrates for exercise: dietary demands for optimal performance. *Int J Sports Med* 1988; 9: 1–18.

(40) Costill, D. L. Gastric emptying of fluids during exercise. In Gisolfi, C. V., Lamb, D. R. (Eds), *Perspectives in Exercise Science and Sports Medicine, Vol. 3, Fluid Homeostatis During Exercise.* Benchmark Press, Carmel, Indiana, 1990, pp. 97–127.

(41) Couzy, F., Lafargue, P., Guezennec, C. Y. Zinc metabolism in the athlete: influence of training, nutrition and other factors. *Int J Sports Med* 1990; 11: 263–266.

(42) Coyle, E. F., Hamilton, M. Fluid replacement during exercise: effects on physiological homeostasis and performance. In Gisolfi, C. V., Lamb, D. R. (Eds), *Perspectives in Exercise Science and Sports Medicine, Vol. 3, Fluid Homeostatis During Exercise.* Benchmark Press, Carmel, Indiana 1990, pp. 281–308.

(43) Coyle, E. F., Coggan, A. R. Effectiveness of carbohydrate feeding in delaying fatigue during prolonged exercise. *Sports Med* 1984; 1: 446–458.

(44) Coyle, E. F. Carbohydrate feedings: effects on metabolism, performance and recovery. In Brouns, F., Saris, W. H. M., Newsholme, E. A. (Eds), *Advances in Nutrition and Top Sport, Med Sport Sci*, Vol. 32. Karger, Basel, 1991, pp. 1–14.

(45) Coyle, E. F., Timing and method of increased carbohydrate intake to cope with heavy training, competition and recovery. *J Sports Sci* 1991; 9: Special Issue, pp. 29–52.

(46) Corley, G., Demarest-Litchford, M., Bazzarre, T. L. Nutrition knowledge and dietary practices of college coaches. *Am Diet Assoc* 1990; 90: 705–709.

(47) Crapo, P. A. Simple versus complex carbohydrate use in the diabetic diet. *Ann Rev Nutr* 1985; 5: 95–114.

(48) Décombaz, J., Arnaud, M. J., Milon, H. *et al.* Energy metabolism of medium chain triglycerides versus carbohydrates during exercise. *Eur J Appl Physiol* 1983; 52: 9–14.

(49) Décombaz, J., Sartori, D., Arnaud, M. J. *et al.* Oxidation and metabolic effects of fructose and glucose ingested before exercise. *Int J Sports Med* 1985; 6: 282–286.

(50) Décombaz, J., Gmuender, B., Sierro, G., Cerretelli, P. Muscle carnitine after strenuous endurance exercise. *J Appl Physiol* 1992; 72(2): 423–427.

(51) Demopoulous, H., Santomier, J., Seligman, M., *et al.* Free radical pathology: Rationale and toxicology of antioxidants and other supplements in sports medicine and exercise science. In

Katch F. (Ed.), *Sport Health and Nutrition*, Human Kinetics Publishers, Champaign, Illinois, 1986.

(52) Deutsche Gesellschaft für Ernährung. *Empfehlungen für die Nährstoffzufuhr*. Umschau Verlag, Frankfurt, 1991.

(53) Dreher, M. L., Dreher, C. J., Berry, J. W. Starch digestibility of foods: a nutritional perspective. *CRC Critical Rev* 1984; 20: 47–71.

(54) Dunnet, W., Crossen, D. The bee pollen promise. *Runners's World* 1980; 15(8): 53–54.

(55) Dwyer, J. Nutritional status and alternative life-style diets with special reference to vegetarism in the U.S. In Rechcigl jr M. (Ed.) *CRC Handbook of Nutritional Supplements, volume 1, Human Use*. CRC Press, New York, 1983.

(56) Eichner, E. R. The anemias of athletes. *Phys Sportsmed* 1986; 14(9): 122–130.

(57) Eichner E. R. Other medical considerations in prolonged exercise. In Lamb, D. R., Murray, R. (Eds), *Perspectives in Exercise Science and Sports Medicine Vol. 1, Prolonged Exercise*. Benchmark Press, Indianapolis, Indiana, 1988, pp. 415–442.

(58) Erp-Baart van, A. M. J., Saris, W. H. M., Binkhorst, R. A. *et al.* Nationwide survey on nutritional habits in elite athletes. Part 1. Energy, carbohydrate, protein and fat intake. *Int J Sports Med* 1989; 10 (Suppl 1): S3–S10.

(59) Erp-Baart van, A. M. J., Saris, W. H. M., Binkhorst, R. A. *et al.* Nationwide survey on nutritional habits in elite athletes. Part II. Mineral and vitamin intake. *Int J Sports Med* 1989; 10 (Suppl 1): S11–S16.

(60) Erp-Baart van, A. M. J. Food habits in athletes. Dissertation, University of Nijmegen, The Netherlands, 1992. ISBN 90-9004526.0.

(61) Evans, G. W. The effect of chromium picolinate on insulin controlled parameters in humans. *Int J Biosoc Med Res* 1989; 11(2): 163–180.

(62) Faber, M., Benadé, A. J. S., van Eck, M. Dietary intake, anthropometric measurements, and blood lipid values in weight training athletes (body builders). *Int J Sports Med* 1986; 7(6): 342–346.

(63) Felig, P., Wahren, J. Fuel homeostasis in exercise. *N Engl J Med* 1975; 293: 1078–1084.

(64) Felig, P. Amino acid metabolism in exercise. *Ann NY Acad Sci* 1977; 301: 56–63.

(65) Felig, P. Hypoglycemia during prolonged exercise in normal man. *N Engl J Med* 1982; 306: 895–910.

(66) Fern, E. B., Bielinski, R. N., Schutz, Y. Effects of exaggerated amino acid and protein supply in man. *Experientia* 1991; 47: 168–172.

(67) Galiano, F. *et al.* Physiological, endocrine and performance effects of adding branched chain amino acids to a 6% carbohydrate–electrolyte beverage during prolonged cycling. *Med Sci Sports Exerc* 1991; 23: S14 (abstract).

(68) Gerster, H. The role of vitamin C in athletic performance. *J Am Coll Nutr* 1989, 8(6): 363–643.

(69) Gisolfi, C. V., Summers, R. W., Schedl, H. P. Intestinal absorption of fluids during rest and exercise. In Gisolfi, C. V., Lamb, D. R. (Eds), *Perspectives in Exercise Science and Sports Medicine, Vol. 3, Fluid Homeostatis During Exercise*. Benchmark Press, Carmel, Indiana, 1990, pp. 129–180.

(70) Gledhill, N. Bicarbonate ingestion and anaerobic performance. *Sports Med* 1984; 1: 177–180.

(71) Gollnick, P. D. Energy metabolism and prolonged exercise. In: Lamb, D. R., Murray, R. (Eds), *Perspectives in Exercise Science and Sports Medicine, Vol. 1, Prolonged Exercise*. Benchmark Press, Indianapolis, Indiana 1988, pp. 1–42.

(72) Grandjean, A. C. The vegetarian athlete. *Phys Sportsmed* 1987; 15(5): 191–194.

(73) Greig, C., Finch, K., Jones, D., *et al.* The effect of oral supplementation with L-carnitine on maximum and sub-maximum exercise capacity. *Eur J Appl Physiol* 1987; 56: 457–460.

(74) Grossman, S. B. (Ed.). *Thirst and Sodium Appetite: Physiological Basis*. Academic Press, New York, 1990.

(75) Guezennec, C. Y., Sabatin, P., Duforez, F. *et al.* Oxidation of corn starch, glucose and fructose ingested before exercise. *Med Sci Sports Exerc* 1989; 21: 45–50.

(76) Guezennec, C. Y., Sabatin, P., Duforez, F. *et al.* Role of starchy food type and structure on the metabolic response to physical exercise. *Med Sci Sports Exerc* 1991 (in press).

(77) Guezennec, C. Y., Leger, C., Sabatin, P. Lipid metabolism and performance. In Atlan, C., Béliveau, L., Bouissou, P. (Eds), *Muscle Fatigue: Biochemical and Physiological Aspects*. Masson, Paris, 1991, pp. 165–172.

(78) Hackman, R. M., Keen, C. L. Changes in serum zinc and copper levels after zinc supplementation in running and non-running men. In Katch, F. I. (Ed.), *Sport, Health and Nutrition. 1984 Olympic Sci Cong Proc*, Vol. 2. Human Kinetics Publishers, Champaign, Illinois, 1986, pp. 89–100.

(79) Hargreaves, M. Carbohydrates and exercise. *J Sport Sci* 1991; 9: Special Issue, pp. 18–28.

(80) Harper, A. E., Zapalowski, C. Metabolism of branched chain amino acids. In Waterlow, J. C., Stephen, J. M. L. (Eds), *Nitrogen Metabolism in Man*. Applied Science Publishers, London, 1981, pp. 97–116.

(81) Hawkins, C. *et al.* Oral arginine does not affect body composition or muscle function in male weight lifters. *Med Sci Sports Exerc* 1991; 23: S15 (abstract).

(82) Hawley, J. A., Dennis, S. C., Noakes, T. D. Oxidation of carbohydrate ingested during prolonged endurance exercise. *Sports Med* 1992, (in press).

(83) Haymes, E. M. The use of vitamin and mineral supplements by athletes. *J Drugs Issues* 1980; 3: 361–370.

(84) Haymes, E. M. Vitamin and mineral supplementation to athletes. *Int J Sport Nutr* 1991; 1: 146–169.

(85) Heigenhauser, G., Jones, N. Bicarbonate loading. In: Lamb, D., Williams, M. (Eds), *Ergogenics; Enhancement of Performance in Exercise and Sport*. Brown & Benchmark, Dubuque, Iowa, 1991.

(86) Herbert, V., Colman, N., Jacob, E. Folic acid and vitamin B12. In Goodhart, R., Shils, M. (Eds), *Modern Nutrition in Health and Disease*. Lea and Febiger, Philadelphia, 1980, pp. 229–258.

(87) Heany, R. P., Recher, R. R., Saville, P. D. Menopausal changes in calcium balance performance. *J Lab Clin Med* 1978; 92: 953–963.

(88) Hubbard, R. W., Szlyk, P. C., Armstrong, L. E. Influence of thirst and fluid palatability on fluid ingestion during exercise. In Gisolfi, C. V., Lamb, D. R. (Eds), *Perspectives in Exercise Science and Sports Medicine, Vol. 3, Fluid Homeostatis During Exercise*, Benchmark Press, Carmel, Indiana, 1990, pp. 39–96.

(89) Hultman, E. Studies on metabolism of glycogen and active phosphate in man with special reference to exercise and diet. *Scand J Clin Lab Invest* 1967; 19: Suppl 94.

(90) Hultman, E. Dietary manipulations as an aid to preparation for competition. In *Proceedings of the World Conference on Sportsmedicine*. Melbourne, 1974, pp. 239–265.

(91) Hultman, E. Liver glycogen in man: effect of different diets and muscular exercise. In Pernow, B., Saltin, B. (Eds), *Muscle Metabolism During Exercise*. New York, Plenum, 1981, pp. 143–152.

(92) Ivy, J. L., Costill, D. L., Fink, W. *et al.* Contribution of medium and long chain triglyceride intake to energy metabolism during prolonged exercise. *Int J Sports Med* 1980; 1: 15–20.

(93) Jacob, R. A., Harold, H., Sandstead, M. D. *et al.* Whole body surface loss of trace metals in normal males. *Am J Clin Nutr* 1981; 34: 1379–1383.

(94) Jacobson, B. Effect of amino acids on growth hormone release. *Phys Sportsmed* 1990; 18: 63–70.

(95) Janssen, G. M. E., Scholte, H. R., Vaandrager-Verduin, M. H. M., Ross, J. D. Muscle carnitine level in endurance training and running a marathon. *Int J Sports Med* 1989; 10: S153–S155.

(96) Keen, C. L., Hackman, R. M. Trace elements in athletic performance. In Katch F. I. (Ed.), *Sport, health and nutrition. 1984 Olympic Sci Cong Proc*, Vol. 2. Human Kinetics Publishers, Champaign, Illinois, 1986, pp. 51–66.

(97) Kieffer F. Trace elements: their importance for health and physical performance. *Deutsche Zschr für Sportmed* 1986; 37: 118–123 (German)

(98) Kiens, B., Raben, A. B., Valeur, A. K., Richter, E. Benefit of dietary simple carbohydrates on the early post exercise glycogen repletion in male athletes. *Med Sci Sports Exerc* 1990; 22: 588 (abstract).

(99) Kreider R. *et al.* Effects of amino acid supplementation on substrate usage during ultraendurance triathlon performance. *Med Sci Sports Exerc* 1991; 23 S16 (abstract).

(100) Kreider, R., Williams M. Phosphate loading and exercise performance. *J Appl Nutr* 1991 (in press).

(101) Lamb, D. R. and A. C. Snyder. Muscle glycogen loading with a liquid carbohydrate supplement. *Int J Sports Nutr* 1991; 1: 52–60.

(102) Lampe, J. W., Slavin, J. L., Apple, F. S. Elevated serum ferritin concentrations in master runners after a marathon race. *Int J Vitamin Nutr Res* 1986; 6: 395–398.

(103) Lane, H. W. Some trace elements related to physical activity: zinc, copper, selenium, chromium and iodine. In Hickson, J. E., Wolinski, I., (Eds), *Nutrition in Exercise and Sport*. CRC Press, Florida, 1989, pp. 301–307.

(104) Lau, K. Phosphate disorders. In Kokko, J. P., Tannen, R. L. (Eds), *Fluids and Electrolyte*. WB Saunders, Philadelphia, 1986, pp. 398–471.

(105) Lemon, P. W. R. Nutrition for muscular development of young athletes. In Gisolfi, C. V., Lamb, D. R. (Eds), *Perspectives in Exercise Science and Sports Medicine, Vol 2, Youth, Exercise, and Sport*. Benchmark Press, Indianapolis, Indiana, 1989, pp. 369–400.

(106) Lemon, P. W. R. Effect of exercise on protein requirements. *J Sports Sci* 1991; 9: Special Issue, pp. 53–70.

(107) Lemon, P. W. R. Does exercise alter dietary protein requirements? In Brouns, F., Saris, W. H. M., Newsholme, E. A. (Eds), *Advances in Nutrition and Top Sport. Med Sport Sci*, Vol. 32. Karger, Basel, 1991, pp. 15–37.

(108) Lemon, P. W. R. Protein and amino acid needs of the strength athlete. *Int J Sport Nutr* 1991; 1: 127–145.

(109) Levander, O. A., Cheng, L. Micronutrient interactions; vitamins, minerals and hazardous elements. *Ann NY Acad Sci*, 1980; 355: 1–372.

(110) Mahalko, J. R., Sandsted, H. H., Johnson, L. K., Milne, D. B. Effect of a moderate increase in dietary protein on the retention and excretion of Ca, Cu, Fe, Mg, P and Zn by adult males. *Am J Clin Nutr* 1983; 37: 8–14.

(111) Massicotte, D., Péronnet, F., Brisson, G., Hillaire-Marcel, C. Exogenous 13 C lipids and 13 C glucose oxidized during prolonged exercise in man. *Med Sci Sports Exerc* 1990; 2: S52, abstr 310.

(112) Massicotte, D., Péronnet, F., Brisson G. *et al*. Oxidation of glucose polymer during exercise: comparison with glucose or fructose. *J Appl Physiol* 1989; 66: 179–183.

(113) Maughan, R. J. Effects of diet composition on the performance of high intensity exercise. In Monod, H. (Ed.), *Nutrition et Sport* 1990: pp. 201–211.

(114) Maughan, R. J. Fluid and electrolyte loss and replacement in exercise. *J Sports Sci* 1991; 9: Special Issue, pp. 117–142.

(115) Maughan, R. J. and Noakes, T. D. Fluid replacement and exercise stress, a brief review of studies on fluid replacement and some guidelines for the athlete. *Sports Med* 1991; 1: 16–31.

(116) Maughan, R. J., Sadler, D. The effects of oral administration of salts of aspartic acid on the metabolic response to prolonged exhausting exercise in man. *Int J Sports Med* 1983; 4: 119–123.

(117) Maughan, R. J., Greenhaff, P. L. High intensity exercise performance and acid–base balance: The influence of diet and induced metabolic alkalosis. In Brouns, F. (Ed.), *Advances in Nutrition and Top Sport. Med Sport Sci*, Vol. 32. Karger, Basel, 1991, pp. 147–165.

(118) McDonald, R., Keen, C. L. Iron, zinc and magnesium nutrition and athletic performance. *Sports Med* 1988; 5: 171–184.

(119) McGilvery R. W. The use of fuels for muscular work. In Howald, H., Poortmans, J. R. (Eds), *Metabolic Adaptation to Prolonged Physical Exercise*. Proceedings of the Second International

Symposium on Biochemistry. Birkhäuser Verlag, Basel. 1973, pp. 12–20.

(120) McNaughton, L. R. Sodium citrate and anaerobic performance: implications of dosage. *Eur J Appl Physiol* 1990; 61: 392–397.

(121) Medbo, J. I., Serjested O. Plasma potassium changes with high intensity exercise. *J Physiol* 1990; 421: 105–122.

(122) Mitchell, M. *et al*. Effects of amino acid supplementation on metabolic responses to ultraendurance triathlon performance. *Med Sci Sports Exerc* 1991; 23: S15 (abstract).

(123) Moses, F. The effect of exercise on the gastrointestinal tract. *Sports Med* 1990; 9: 159–172.

(124) Mosora, F., Lacroix, M., Luyckx, A. S. *et al*. Glucose oxidation in relation to the size of the oral glucose loading dose. *Metabolism* 1981; 30: 1143–1149.

(125) Munro, H. N. Metabolism and functions of amino acids in man—overview and synthesis. In Blackburn, G. L., Grant, J. P., Vernon, R. Y. (Eds), *Amino Acids, Metabolism and Medical Applications*. John Wright, PSG, Boston, 1983, pp. 1–12.

(126) Murray, R. The effects of consuming carbohydrate–electrolyte beverages on gastric emptying and fluid absorption during and following exercise. *Sports Med* 1987; 4: 322–351.

(127) Murray, R., Seifert, J. G., Eddy, D. E. *et al*. Carbohydrate feeding and exercise: effect of beverage carbohydrate content. *Eur J Appl Physiol* 1989; 59: 152–158.

(128) Murray, R., Paul, L. P., Seifert, J. G. *et al*. The effects of glucose, fructose and sucrose ingestion during exercise. *Med Sci Sports Exerc* 1989; 3: 275–282.

(129) Nadel, E. R. Temperature regulation and prolonged exercise. In Lamb, D. R., Murray, R. (Eds), *Perspectives in Exercise Science and Sports Medicine, Vol. 1, Prolonged Exercise*. Benchmark Press, Indianapolis, Indiana, 1988, pp. 125–151.

(130) Nadel, E. R., Mack, G. W., Nose H. Influence of fluid replacement beverages on body fluid homeostasis during exercise and recovery. In Gisolfi, C. V., Lamb, D. R. (Eds), *Perspectives in Exercise Science and Sports Medicine, Vol. 3, Fluid Homeostatis During Exercise*, Benchmark Press, Carmel, Indiana, 1990, pp. 181–205.

(131) National Research Council. *Recommended Dietary Allowances*, 1989; 10th edition, Washington, National Academy Press.

(132) Newhouse, I. J., Clement, D. B. Iron status in athletes. An update. *Sports Med* 1988; 5: 337–352.

(133) Newsholme, E. A., Start, C. (Eds). Regulation of glycogen metabolism. In *Regulation in Metabolism*. John Wiley & Sons, Chichester, 1973, pp. 146–194.

(134) Newsholme, E. A., Start, C. (Eds). Adipose tissue and the regulation of fat metabolism. In *Regulation in Metabolism*. John Wiley & Sons, Chichester, 1973, pp. 195–246.

(135) Newsholme, E. A., Start, C. (Eds). Regulation of carbohydrate metabolism in liver. In *Regulation in Metabolism*. John Wiley & Sons, Chichester, 1973, pp. 247–323.

(136) Newsholme, E. A., Leech A. R. (Eds). Integration of carbohydrate and lipid metabolism. In *Biochemistry for the Medical Sciences*. John Wiley & Sons, Chichester, 1983, pp. 336–356.

(137) Newsholme, E. A., Leech, A. R. (Eds). Metabolism in exercise. In *Biochemistry for the Medical Sciences*. John Wiley & Sons, Chichester, 1983, pp. 357–381.

(138) Newsholme, E. A., Leech, A. R. (eds). The integration of metabolism during starvation, refeeding, and injury. In *Biochemistry for the Medical Sciences*. John Wiley & Sons, Chichester, 1983 pp. 536–561.

(139) Newsholme, E. A., Parry-Billings, M., McAndrew N. *et al*. A biochemical mechanism to explain some characteristics of overtraining. In Brouns, F., Saris, W. H. M., Newsholme, E. A. (Eds), *Advances in Nutrition and Top Sport. Med Sport Sci*, Vol. 32. Karger, Basel, 1991, pp. 79–93.

(140) Newsholme, E. Effects of exercise on aspects of carbohydrate, fat, and amino acid metabolism. In Bouchard C., Shephard R., Stephens T. *et al*. (Eds). *Exercise, Fitness and Health*. Human Kinetics, Publishers Champaign, Illinois, 1990.

(141) Nieman, D. C. Vegetarian dietary practices and endurance performance. *Am J Clin Nutr* 1988; 48: 754–761.

(142) Noakes, T. D., Goodwin, N., Rayner, B. L. *et al*. Water intoxication: a possible complication during endurance exercise. *Med Sci Sports Exerc* 1985; 17: 370–375.

(143) Noakes, T. D., Adams, B. A., Myburgh, K. H. *et al*. The danger of an inadequate water intake during prolonged exercise. A novel concept revisited. *Eur J Appl Physiol* 1988; 57: 210–219.

(144) Noakes, T. D., Normann, R. J., Buck, R. H. *et al*. The incidence of hyponatremia during prolonged ultraendurance exercise. *Med Sci Sports Exerc* 1990; 22: 165–170.

(145) Oppenheimer, S., Hendrickse, R. The clinical effects of iron deficiency and iron supplementation. *Nutr Abs Rev* 1983; 53: 585–598, series A.

(146) Otto, R., Shores, K., Wygard, J. *et al*. The effects of L-carnitine supplementation on endurance exercise. *Med Sci Sports Exerc* 1987; 19: S87 (abstract).

(147) Oyono-Enguelle, S., Freund, H., Ott, C. *et al.* Prolonged submaximal exercise and L-carnitine in humans. *Eur J Appl Physiol* 1988; 58: 53–61.

(148) Pallikarakis, N., Jandrain, B., Pirnay, F. *et al.* Remarkable metabolic availability of oral glucose during long duration exercise in humans. *J Appl Physiol* 1986; 60: 1035–1042.

(149) Parr, R. B., Porter, M. A., Hodgson, S. C. Nutrition knowledge and practice of coaches, trainers and athletes. *Phys Sportsmed* 1984; 12: 127–138.

(150) Pate, R. R. Sports anemia: a review of the current research literature. *Phys Sportsmed* 1983; 11: 115–131.

(151) Pate, R. R., Sargent, R. G., Baldwin, C., Burgess, M. L. Dietary intake of women runners. *Int J Sports Med* 1990; 11(6): 461–466.

(152) Puhl, J. L., Handel van, P. J., Williams, L. L. *et al.* Iron status and training. In Butts, N. K., Gushiken, T. T., Zarins, B. (Eds), *The Elite Athlete*. Life Enhancement Publications, Champaign, Illinois, 1985, pp. 209–238.

(153) Rehrer, N. J., Beckers, E., Brouns, F. *et al.* Exercise and training effects on gastric emptying of carbohydrate beverages. *Med Sci Sports Exerc* 1989; 21: 540–549.

(154) Rehrer, N. J. *et al.* Gastric emptying, secretion and electrolyte flux after ingestion of beverages with varying electrolyte compositions. PhD Thesis 1990, University of Limburg, Maastricht, The Netherland, *Med Sci Sports Exerc* 1991 (in press).

(155) Rehrer, N. J. Aspects of dehydration and rehydration during exercise. In Brouns, F., Saris, W. H. M., Newsholme, E. A. (Eds), *Advances in Nutrition and Top Sport. Med Sport Sci*, Vol. 32. Karger, Basel, 1991, pp. 128–146.

(156) Rehrer, N. J., van Kemenade, M. C., Meester, T. A., Brouns F., Saris W. H. M. Gastroinstestinal complaints in relation to dietary intakes in triathletes. *Int J Sport Nutr* 1992 (in press).

(157) Rennie, M. J., Edwards, R. H. T., Halliday, D. *et al.* Protein metabolism during exercise. In Waterlow, J. C., Stephen, J. M. L. (Eds), *Nitrogen Metabolism in Man*. Applied Science Publishers, London, 1981, pp. 509–524.

(158) Richter, E. A., Sonne, B., Plough, T. *et al.* Regulation of carbohydrate metabolism in exercise. In Saltin, B (Ed.), *Biochemistry of Exercise 6*. Human Kinetics Publishers, Champaign, Illinois, 1986; pp. 151–166.

(159) Roberts, J. The effect of coenzyme Q10 on exercise performance. *Med Sci Sports Exerc* 1990; 22: S87 (abstract).

(160) Rumessen, J. J., Gudmand-Hoyer, E. Absorption capacity of fructose in healthy adults. Comparison with sucrose and its consistent monosaccharides. *Gut* 1986; 27: 1161–1168.

(161) Russel, McR. D., Jeejeeboy, K. N. The assessment of the functional consequences of malnutrition. *Nutr Abstr and Rev in Clin Nutr, series A* 1983; 53 (10): 863–877.

(162) Sabatin, P., Portero, P., Gilles, D. *et al.* Metabolic and hormonal responses to lipid and carbohydrate diets during exercise in man. *Med Sci Sports Exerc* 1987; 19(3): 218–223.

(163) Saris, W. H. M., Brouns, F. Nutritional concerns for the young athlete. In Rutenfranz, J., Mocellin, R., Klinit, F. (Eds), *International Series on Sports Sciences. Vol. 17, Children and exercise XII.* Human Kinetics Publishers, Champaign, Illinois, 1986, pp. 11–18.

(164) Saris, W. H. M., Schrijver, J., Erp Baart van, M. A., Brouns, F. Adequacy of vitamin supply under maximal sustained workloads: The Tour de France. In Walter, P., Brubacher, G. B., Stähelin, H. B. (Eds), *Elevated Dosages of Vitamins.* Huber Publishers, Toronto, 1989.

(165) Saris, W. H. M., van Erp Baart, M. A., Brouns, F. *et al.* Study on food intake and energy expenditure during extreme sustained exercise: The Tour de France. *Int J Sports Med* 1989; 10: S26–S31 (Suppl).

(166) Sawka, M. N. Body fluid responses and hypohydration during exercise–heat stress. In Pandolf, K. B., Sawka, M. N., Gonzalez, R. R. (Eds), *Human Performance Physiology and Environmental Medicine at Terrestrial Extremes.* Benchmark Press, Indianapolis, Indiana, 1988, pp. 227–266.

(167) Sawka, M. N., Wenger C. B. Physiological responses to acute exercise–heat stress. In Pandolf, K. B., Sawka, M. N., Gonzalez, R. R. (Eds), *Human Performance Physiology and Environmental Medicine at Terrestrial Extremes.* Benchmark Press, Indianapolis, Indiana, 1988, pp. 97–151.

(168) Sawka, M. N., Pandolf, K. B. Effects of body water loss on physiological function and exercise performance. In Gisolfi, C. V., Lamb, D. R. (Eds), *Perspectives in Exercise Science and Sports Medicine, Vol. 3, Fluid Homeostatis During Exercise.* Benchmark Press, Carmel, Indiana, 1990, pp. 1–38.

(169) Schoutens, A., Laurent, E., Poortmans, J. R. Effect of inactivity and exercise on bone. *Sports Med* 1989; 7: 71–81.

(170) Segura, R., Ventura, J. Effect of L-tryptophan supplementation on exercise performance. *Int J Sports Med* 1988; 9: 301–305.

(171) Shephard, R. J. Vitamin E and athletic performance. *J Sports Med* 1983; 23: 461–470.

(172) Sherman, W. M. Carbohydrates, muscle glycogen, and muscle glycogen supercompensation. In Williams, M. H. (Ed.), *Ergogenic Aids in Sport*. Human Kinetics Publishers, Champaign, Illinois, 1983, pp. 3–26.

(173) Sherman, W. M., Lamb, D. R. Nutrition and prolonged exercise. In Lamb, D. R., Murray, R. (Eds), *Perspectives in Exercise Science and Sports Medicine, Vol. 1, Prolonged Exercise*. Benchmark Press, Indianapolis, Indiana 1988, pp. 213–280.

(174) Sherman, W. M., Wimer, G. S. Insufficient dietary carbohydrate during training: does it impair performance. *Int J Sports Nutr* 1991; 1: 28–44.

(175) Shores, K., Otto, R., Wygard, J. *et al*. Effect of L-carnitine supplementation on maximal oxygen consumption and free fatty acid serum levels. *Med Sci Sports Exerc* 1987; 19: S60 (abstract).

(176) Short, S. H., Short, W. R. Four-year study of university athletes' dietary intake. *Am Diet Assoc* 1983; 82(6): 632–645.

(177) Smith, J. C., Morris, E. R., Ellis, R. Zinc requirements, bioavailabilities and recommended dietary allowances. In Prasad, A. A. (Eds), *Zinc Deficiency in Human Subjects*. Alan R Liss, New York, 1983, pp. 147–169.

(178) Spencer, H., Kramer, L., Perakis, E. *et al*. Plasma levels of zinc during starvation. *Fed Proc* 1982; 41: 347.

(179) Steben, R., Wells, J., Harless, I. The effects of bee pollen tablets on the improvement of certain blood factors and performance of male collegiate swimmers. *J Nat Athletic Trainers Assoc* 1976; 11: 124–126.

(180) Storlie, J. Nutrition assessment of athletes: A model for integrating nutrition and physical performance indicators. *Int J Sport Nutr* 1991; 1: 192–204.

(181) Sutton, J. R. Clinical implications of fluid imbalance. In Gisolfi, C. V., Lamb, D. R. (Eds), *Perspectives in Exercise Science and Sports Medicine, Vol. 3, Fluid Homeostatis During Exercise*. Benchmark Press, Carmel, Indiana, 1990, pp. 425–455.

(182) Tarnopolsky, M. A., MacDougall, J. D., Atkinson, S. A. Influence of protein intake and training status on nitrogen balance and lean body mass. *J Appl Physiol* 1988; 64(1): 187–193.

(183) Trichopoulou, A., Vassilakos, T. Recommended dietary intakes in the European Community member states. *Eur J Clin Nutr* 1990; Suppl 2: 51–101.

(184) Vandewalle, L. *et al*. Effect of branched-chain amino acid supplements on exercise performance in glycogen depleted subjects. *Med Sci Sports Exerc* 1991; 23: S116, (abstract).

(185) Vollestad, N. K, Serjested, O. M. Plasma K$^+$ shifts in muscle and blood during and after exercise, *Int J Sports Med* 1989; 10 (Suppl 2): 101.

(186) Wade, C. E., Freund, B. J. Hormonal control of blood volume during and following exercise. In Gisolfi, C. V., Lamb, D. R. (Eds), *Perspectives in Exercise Science and Sports Medicine, Vol. 3, Fluid Homeostatis During Exercise*. Benchmark Press, Carmel, Indiana, 1990, pp. 207–245.

(187) Wagenmakers, A. J. M., Coakley, J. H., Edwards, R. H. T. Metabolism of branched-chain amino acids and ammonia during exercise: clues from McArdle's disease. *Int J Sports Med* 1990; 11 (Suppl. 2): S101–S113.

(188) Wagenmakers, A. J. M., Beckers, E. J., Brouns, F. *et al.* Carbohydrate supplementation, glycogen depletion, and amino acid metabolism during exercise. *Am J Physiol* 1991; 260 (endocrinol. metab. 23): E883–E890.

(189) Wagenmakers, A. J. M. A role of amino acids and ammonia in mechanisms of fatigue. In Marconnet, P., Saltin, B., Komi, P. (Eds), *Local Fatigue in Exercise and Training, 4th Int Symposium on Exercise and Sport Biology*, Nice 1991. *Med Sport Sci*, **34**, (1992, in press).

(190) Wagenmakers, A. J. M. L-Carnitine supplementation and performance in man. In Brouns, F. (Ed.), *Advances in Nutrition and Top Sport*, Vol. 32, Karger, Basel, 1991, pp. 110–127.

(191) Walberg, J. L., Leidy, M. K., Sturgill, D. J. *et al.* Macronutrient content of a hypoenergy diet affects nitrogen retention and muscle function in weight lifters. *Int J Sports Med* 1988; 9(4): 261–266.

(192) Warren, B. *et al.* The effect of amino acid supplementation on physiological responses of elite junior weightlifters. *Med Sci Sports Exerc* 1991; 23: S15 (abstract).

(193) Wassermann, D. H., Geer, R. J., Williams, P. E. *et al.* Interaction of gut and liver in nitrogen metabolism during exercise. *Metabolism* 1991; 40: 307–314.

(194) Waterlow, J. C. Metabolic adaptation to low intakes of energy and protein. *Ann Rev Nutr* 1986; 6: 495–526.

(195) Weight, L. M., Myburgh, K. M., Noakes, T. D. Vitamin and mineral supplementation: effect on the running performance of trained athletes. *Am J Clin Nutr* 1988; 47: 192–195.

(196) Wenger C. B. Human heat acclimatization. In Pandolf, K. B., Sawka, M. N., Gonzalez, R. R. (Eds), *Human Performance Physiology and Environmental Medicine at Terrestrial Extremes*. Benchmark Press, Indianapolis, Indiana, 1988, pp. 153–197.

(197) Wesson, M., McNaughton, L., Davies, P. *et al.* Effects of oral administration of aspartic acid salts on the endurance capacity of trained athletes. *Res Quarterly Exerc Sport* 1988; 59: 234–239.

(198) White, S. L., Maloney, S. K. Promoting healthy diets and active lives to hard-to-reach groups: market research study. *Public Health Rep* 1990; 105(3): 224–231.

(199) Williams, M. H. *Nutritional Aspects of Human Physical and Athletic Performance* (2nd edition). Charles C. Thomas, Springfield, Illinois, 1985.

(200) Williams, M., Kreider, R., Hunter, D. *et al.* Effect of oral inosine supplementation on 3-mile treadmill run performance and VO_2 peak. *Med Sci Sports Exerc* 1990; 22: 517–522.

(201) Williams, M. Ergogenic aids. In: Berning, J., Steen, S. (Eds), *Sports Nutrition for the 90s: The Health Professional's Handbook*, Aspen Publishers, Gaithersburg, Maryland 1991.

(202) Williams, M. H. *Nutrition for Fitness and Sport* (3rd edition). Wm. C. Brown Publishers, Dubuque, Iowa, 1992 (in press).

(203) Wilmore, J., Freund, B. J. Nutritional enhancement of physical performance. *Nutrition Abstracts and Reviews in Clinical Nutrition* series A 1984; 54(1): 1–16.

(204) Wolfe, R. R., Richard, D., Goodenough, D. *et al.* Protein dynamics in stress. In Blackburn, G. L., Grant, J. P., Vernon, R. Y. (Eds), *Amino Acids, Metabolism and Medical Applications*. John Wright, PSG, Boston, 1983, pp. 396–413.

(205) Woodhouse, M., Williams, M., Jackson, C. The effects of varying doses of orally ingested bee pollen extract upon selected performance variables. *Athletic Training* 1987; 22: 26–28.

(206) Wretlind, A. Nutrition problems in healthy adults with low activity and low caloric consumption. In Blix, G. (Ed.), *Nutrition and Physical Activity*, 1976, pp. 114–131.

(207) Wurtman, R. J., Lewis, M. C. Exercise, plasma composition and neurotransmission. In Brouns, F., Newsholme, E. A., Saris, W. H. M. (Eds), *Advances in Nutrition and Top Sport. Med Sport Sci*, Vol. 32. Karger, Basel, 1991, pp. 94–109.

(208) Wyss V. *et al.* Effects of L-carnitine administration on VO_2max and the aerobic–anaerobic threshold in normoxia and acute hypoxia. *Eur J Appl Phys* 1990; 60: 1–6.

(209) Yoshimura, H., Inoue, T., Yamada, T., Shiraki, K. Anemia during hard physical training (sports anemia) and its causal mechanisms with special reference to protein nutrition. *Wld Rev Nutr Diet* 1980; 35: 1–86.

(210) Zuliani, U., Bonetti, A., Campana, M., *et al*. The influence of ubiquinone (CoQ10) on the metabolic response to work. *J Sports Med Phys Fitness* 1989; 29: 57–61.

(211) Maughan, R. J. Exercise-induced muscle cramp: a prospective biochemical study in marathon runners. *J Sports Sci* 1986; 4: 31–34.

(212) Brouns, F., Saris, W. H. M., Schneider, H. Rationale for upper limits of electrolyte replacement during exercise. *Int J Sport Nutr* 1992; 2: 229–238.

(213) Wagenmakers, A. J. M., Brouns, F., Saris, W. H. M., Halliday, D. Maximal oxidation of oral carbohydrates during exercise. *Med Sci Sports Exerc* 1990; 22 (4): S120, abstract.

(214) Brouns, F., Beckers, E., Knöpfli, B., Villiger, B., *et al*. Rehydration during exercise: effect of electrolyte supplementation on selective blood parameters. *Med Sci Sports Exerc* 1991; 23 (4): S84, abstract.

(215) Rehrer, N. J., Brouns, F., Beckers, E. J., *et al*. Gastric emptying with repeated drinking during running and cycling. *Int J. Sports Med* 1990; 11: 238–243.

(216) Kazunori, N., Clarkson, P. M. Changes in plasma zinc following high force eccentric exercise. *Int J. Sport Nutr* 1992; 2: 175–184.

Index

carbohydrate (CHO) (*continued*)
  food sources 11, *12*, 21
  intake at rest 26–8
  liberated during muscle work
    14–15
  loading *23*
  oral 20–6
  oxidation during exercise 24–5, *25*
  in rehydration solutions 62–3, 67,
    69–70
  reserves 11–20
  shortage, effect on protein
    requirement 39
  supplementation during exercise
    20–6
  use during endurance event 3–4
  utilization 20, 31–2, 62
carbohydrate–electrolyte
  rehydration solutions 63–70
carbohydrate:fat ratio 14
carbon dioxide 56
cardiomyopathy 94
carnitine 101, 126
L-carnitine 115–16, *116*
ceruloplasmin 91
chloride 54, 56
  in fluid homestasis 71
  minimal daily requirements 60
chromium 92–4, 124, 125
  influence of exercise 93–4
  insulin and 92, 93
  intake 94
  sources 94
chromium picolinate 93
chylomicrons 131
citric acid (Kreb's) cycle 29, 39, 129,
  130, 132–3, 136, 138
coffee 85
collagen 101
collapse 4
copper 91–2, 96, 124
  influence of exercise 91
  intake 92
  recommended dietary intakes 8
  sources 92
  status 91, 92
CoQ10 (ubiquinone) 117
creatine phosphate 14
cyanocobalamin 99

dehydration 4, 51–70

intercellular 55
delayed onset of muscle soreness
  (DOMS) 43
diabetes 92
diarrhea 60, 67
  fluid loss 123
  following sodium bicarbonate 120
  potassium losses 73
dietetic food products 8
diphosphoglycerate 118
disaccharides 24
2,3-DPG (diphosphoglycerate) 118

electrolytes 51–70, 123–4
  extracellular 55–9
  intake 60–70
  intracellular 54–5
endurance exercise capacity 38
endurance performance capacity 82
energy expenditure
  daily 1
  during athletic event 1–3, *2*
  during training days 3
  measurement of *27*
energy metabolism 137–8
enzymatic stimulation tests 97
enzymes 43, 82
esterification 132
estrogen 79
euglycemia 24

fat 28–36, 122–3
  body content 29, *30*
  influence of exercise 31–2, 34
  intake 34–6
  intramuscular 33–4
  liberated during muscle work
    14–15
  mobilization 31–2
  muscle 32–4
  reserves 29–34
  sources 123
  subcutaneous tissue 29
  supplementation 34–6
  transport and utilization *32*
fat:carbohydrate ratio 14
fat droplets 33, *33*, 130
fatigue 24
  ammonia and 39
  central 13
  glycogen and 13

Compiled by Annette Musker